A Guide to the 2006 IRC® Wood Wall Bracing Provisions

APA
THE ENGINEERED WOOD ASSOCIATION

ICC
INTERNATIONAL CODE COUNCIL®

A Guide to the 2006 IRC Wood Wall Bracing Provisions

ISBN: 978-1-58001-623-0

By

International Code Council Inc.
500 New Jersey Avenue, NW, 6th Floor
Washington, DC 20001

and

APA – The Engineered Wood Association
7011 South 19th Street
Tacoma, Washington 98466

The information contained in this guide is believed to be accurate; however, it is being provided for informational purposes only. Publication of this document by the ICC should not be construed as the ICC or the APA engaging in or rendering engineering, legal or other professional services. Use of the information contained in this book should not be considered by the user as a substitute for the advice of a registered professional engineer, attorney or other professional. If such advice is required, it should be sought through the services of a registered professional engineer, licensed attorney or other professional.

First printing: August 2007

Printed in the United States of America

TABLE OF CONTENTS

A GUIDE TO THE 2006 IRC WOOD WALL BRACING PROVISIONS

PREFACE

This illustrated guide was developed to help building officials, designers, and builders apply the bracing requirements of the 2006 International Residential Code® (IRC). While lateral bracing is just one of many important factors to consider when designing, performing plan review, building and inspecting a wood frame structure, it appears to be one of most common sources of confusion and misapplication. To develop the contents of this publication, the authors worked closely with the International Code Council® (ICC) staff to identify and explain the key elements of bracing and to demystify the prescriptive bracing provisions of the IRC.

The most basic concepts of forces acting on the building, the historical perspective and the correct application of the IRC bracing provisions have been presented in a logical format, often incorporating appropriate examples, specific applications and suggested solutions. The final chapter of the book is dedicated to whole house applications where several residential building plans in various wind, snow and seismic regions have been analyzed for proper wall bracing application.

Other aspects of the structure that work in conjunction with bracing to form a safe building, such as diaphragms, wall cladding and inter-element connections are discussed briefly, when necessary to enhance understanding of bracing.

The information contained in this book, though written to explain the prescriptive bracing provisions of the 2006 edition of the IRC, can also be applied to those bracing provisions of the International Building Code® (IBC) Section 2308 (Conventional Light Frame Construction) that are similar to the IRC. The underlying concepts and theory of bracing are identical, but the provisions have evolved differently in the two code documents.

The ultimate goal of this publication is to promote the accurate understanding and the correct application of the code, resulting in safer buildings and communities – a goal that the APA and ICC share.

The authors of the book are three of the wall bracing experts from *APA – The Engineered Wood Association:*

Edward Keith, P.E., is the Manager for Codes and Engineering. With over 20 years experience in wood engineering, product development and building code development, he has served on numerous national committees including the SBCCI Standards for Hurricane Resistant Residential Construction (SSTD-10) and Seismic Resistant Construction (SSTD-13), and is currently a member of the ICC Ad Hoc Committee on Wall Bracing. He is currently registered in the States of Florida and Washington.

Zeno Martin, P.E., is a Senior Engineer. He has many years of experience with wood engineering, testing large-scale wood wall bracing assemblies, and building code development and has earned both a B.S. and an M.S. in civil engineering from Michigan Technological University, with an emphasis in structural engineering. He is currently registered in the States of California and Washington.

Greg Bates is an Engineered Wood Specialist. His responsibilities include promoting the proper application of engineered wood products for efficient and durable building solutions. He received his B.S. degree in Industrial Technology from the University of Wisconsin – Stout.

Contributors to the book are two technical staff of ICC:

John Henry, P.E., Principal Staff Engineer and, **Hamid Naderi, P.E.**, Vice President of Product Development. Their technical review and input, evaluation, analysis, suggestions, clarifications, interpretations, direction, good ideas and good humor were invaluable to the process and the quality of the end product.

Also worth mentioning are the valuable technical reviews by other ICC staff from the Technical Services Department and others who worked behind the scenes to pull this project together.

The technical team would have been lost without the efforts of the APA Market Communications staff who developed all of the figures, edited the text, and designed the pages of this book: Michael Martin, Andrew Sternard, Mary Trodden, Emily Westman, and Marilyn LeMoine.

And last, but certainly not the least, ICC and APA would like to thank the staff from other industry organizations who contributed their valuable time to thoroughly review this publication. Without exception, each reviewer brought valuable insights to the table and added real value to the final product. Our thanks go out to Simpson Strong-Tie, Weyerhaeuser Company, and the American Wood Council. Randy Shackelford, Dave Gromala, Sandra Hyde, Philip Line and other individuals from these companies and organizations provided invaluable input.

CHAPTER 1

Understanding the Lateral Forces that Act on a House

Any building, regardless of size or location, must be designed to safely resist the structural loads anticipated during its lifetime. These loads can be divided into two categories: *vertical loads* and *lateral loads*.

Vertical loads are loads acting in the "up" or "down" direction. In most cases the "down" loads are caused by gravity. These loads are the obvious ones: the weight of the building itself (dead load), the weight of everything in the building (live load), and variable loads, such as those from snow. An example of an "up" load is wind uplift. The "up" load path is the same as the "down"/gravity load path except that the load is acting in the "up" direction. These loads are easy to understand and typical construction practice has evolved into an efficient system that does a good job of accommodating them. Generally speaking, builders in high wind areas are as comfortable installing uplift straps as they are placing beams on cripple studs.

The vertical load path in the upward direction is reflective of the vertical load path in the downward direction. Because downward loads are always present (due to gravity), any deficiencies in the vertical load path are almost immediately apparent due to structural instability. For example, a beam with support at only one end is not going to stand up during construction.

The challenge lies with the "other loads," the lateral loads. Lateral loads are those that act in a direction parallel to the ground. The two major sources of lateral load on a house are wind and seismic (earthquake) forces. Because lateral loads are infrequent, there are no immediate indicators of a complete lateral load path.

In all parts of the country, to one extent or another, lateral load resistance – part of which is wall bracing – has to be built into the structure during construction. While this is especially important in areas of high wind and seismic forces where the lateral force can be very large, the provisions of the International Residential Code (IRC) make it an important consideration for all parts of the country. The IRC prescriptively requires specific building elements to resist lateral forces for all residential structures within its scope. Wood frame construction makes it easy for building professionals to construct strong, durable structures that meet code requirements and assure good performance during severe weather events and earthquakes. It is essential, however, that the structure be properly detailed.

When building to meet the code requirements for seismic or wind in any part of the country, it's important to understand how lateral loads act on wood framing systems, and how construction detailing and fasteners affect the ultimate performance of a structure. Builders, designers and building officials can use these principles to ensure strength, quality, and safety in both residential and non-residential buildings. Certainly, a better understanding of the principles will ensure fewer mistakes in the field.

Wind forces

During a wind event, the wind pushes against one end wall while pulling on the opposite end wall. See Figure 1.1. Because the two walls recieving wind pressures push and pull the roof in the same direction as the wind, the walls on the other two sides of the structure, the bracing walls, must hold the roof from moving. Thus, all walls must be strong enough to resist the wind forces that push against the home.

FIGURE 1.1

WIND FORCES ACTING ON A STRUCTURE

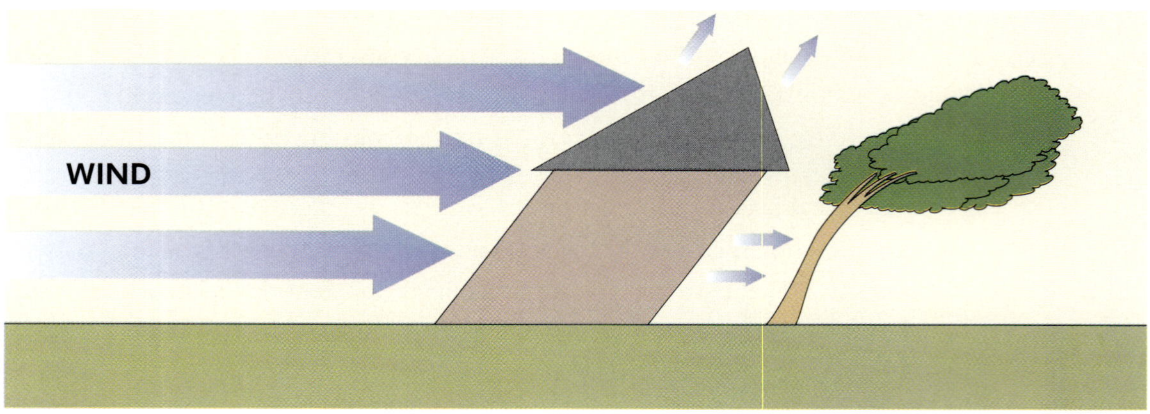

The IRC applies to wind speeds less than 110 mph. In hurricane-prone regions, however, this upper limit is reduced to 100 mph. In areas with wind speeds 110 and over, or 100 and over in hurricane-prone regions, residential structures must be designed in accordance with the International Building Code (IBC), or other referenced standards as discussed in Chapter 4. It should be noted that the IRC bracing provisions are not intended to provide resistance to the extreme wind conditions that occur in the path of tornadoes.

In addition to the wind speed, it is often important to know the Exposure Category of the building site. Wind Exposure Category is explained in Section R301.2.1.4 of the IRC and Chapter 4 of this book. While not necessary for using the prescriptive bracing portions of the code, it is a necessity if parts or portions of the structure fall outside of the prescriptive requirements of the IRC. The Wind Exposure Category is also required to determine component and cladding load requirements of Section R301.2.1.

The IRC contains wind maps for all regions of the country in Section R301.2.1. A sample wind map is shown in Figure 1.2.

FIGURE 1.2
EXAMPLE OF DESIGN WIND SPEED MAP. REPRINT OF IRC FIGURE R301.2(4)
(Wind speeds shown in miles per hour.)

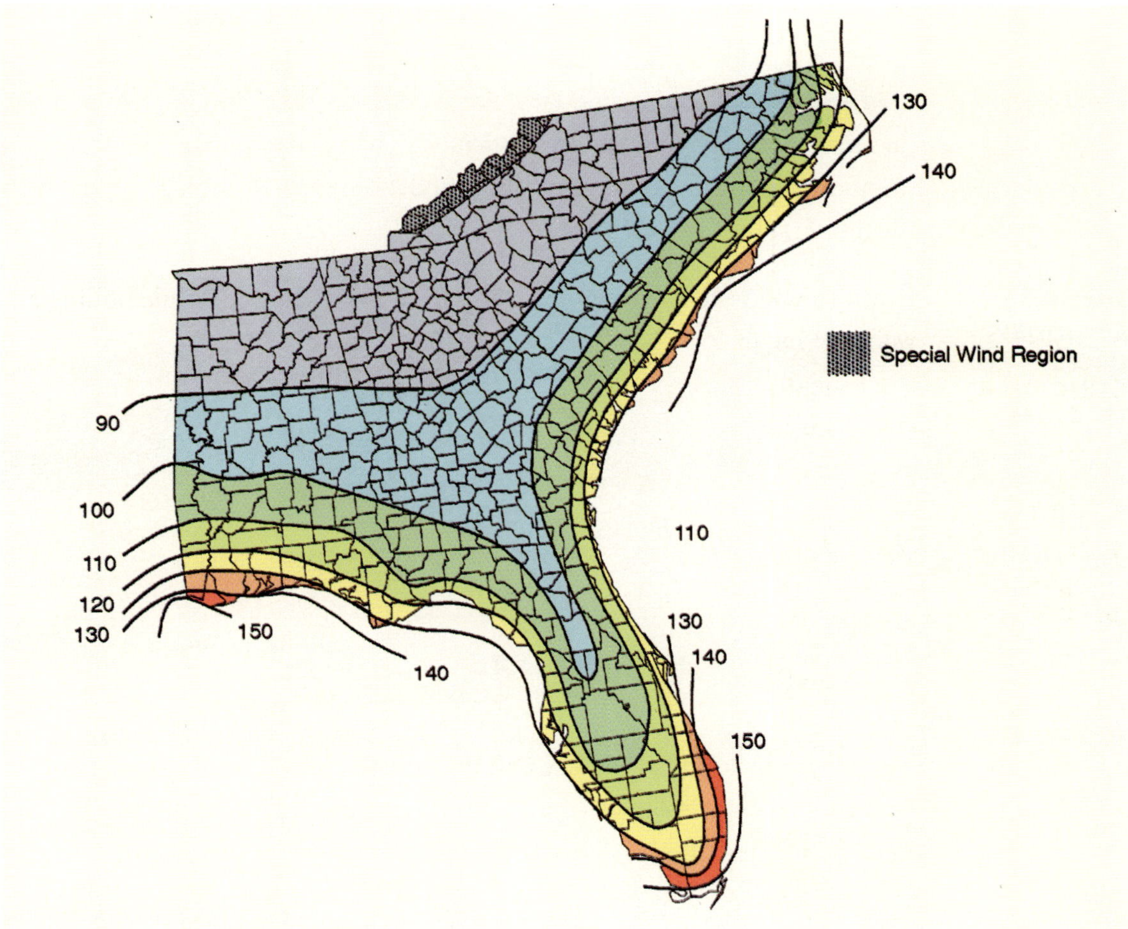

Seismic forces

Seismic forces are generated by ground motions during an earthquake. The ground motion causes the structure's mass to accelerate back and forth, up and down. This acceleration causes forces to develop within the structure in locations where the structure's mass is the largest (remember Newton's Second Law, Force = Mass x Acceleration). Essentially, the seismic ground motion acts on the foundation (*acceleration*), while inertia (*mass* of the structure) attempts to keep the roof from moving with the foundation, causing shear *forces* on the walls. This can be seen in Figure 1.3.

FIGURE 1.3
EARTHQUAKE FORCES ACTING ON A STRUCTURE
(Vertical – upward forces not shown for clarity.)

Similar to the wind maps discussed previously, the IRC publishes earthquake maps, broken down by seismic design categories, for various parts of the country. These maps are located in Section 301.2 of the IRC and a sample map is shown in Figure 1.4.

The use of seismic design categories (SDC) is a simplified means of measuring the hazard of an earthquake in a given region. These categories are measured from A to E, with A being the lowest and D_2 being the highest covered by the IRC. For Seismic Design Category E, the IRC cannot be used and the IBC, or other referenced standards must be used to design the structure.

As can be expected, as the SDC increases from A to D_2, the bracing and attachment requirements for a given structure increase. In addition, as discussed in Chapter 4, the IRC has additional limitations on irregular portions of structures in higher SDCs.

FIGURE 1.4

EXAMPLE OF SEISMIC DESIGN CATEGORY MAP

REPRINT OF IRC FIGURE R301.2(2)

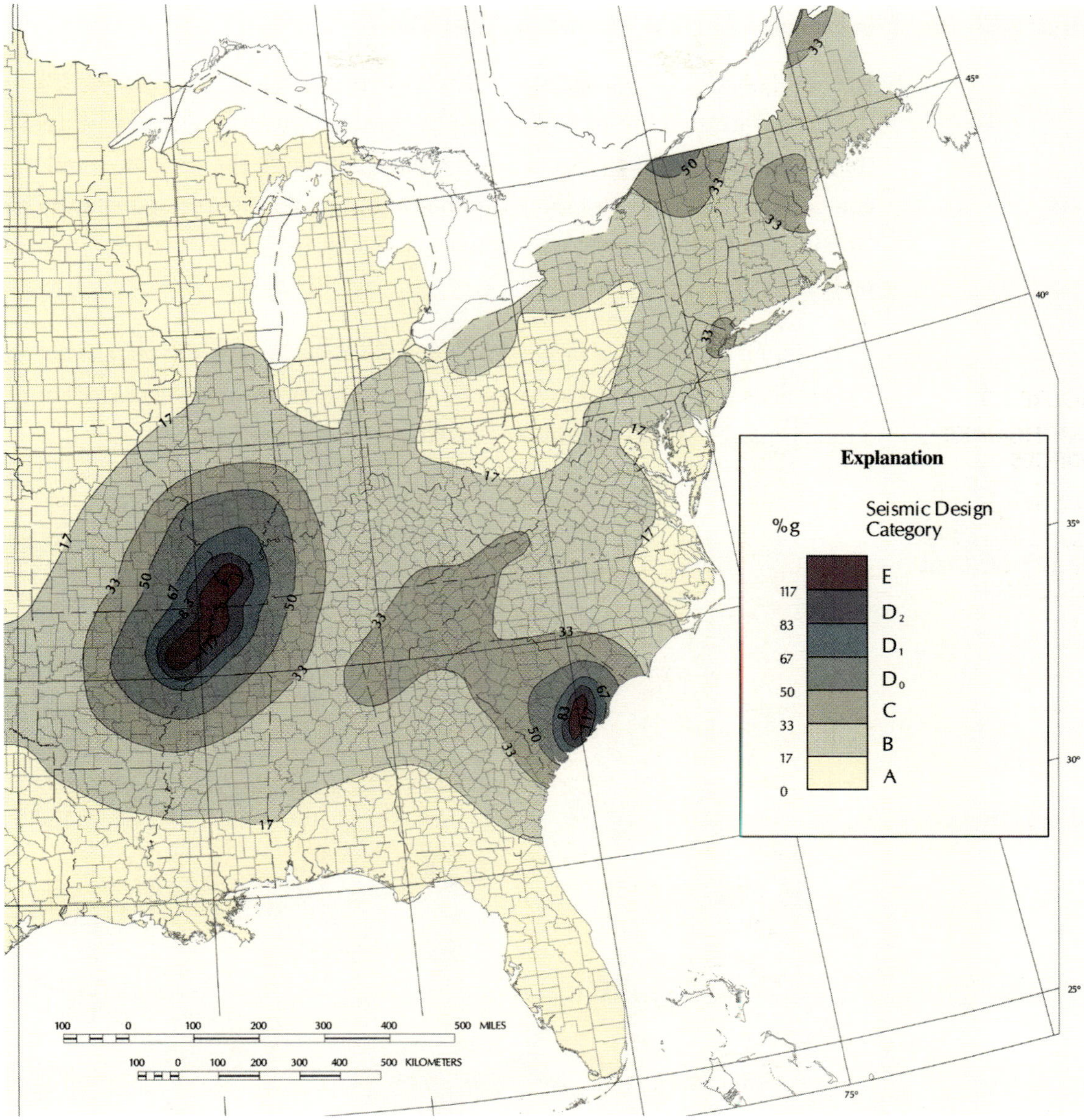

From the wind and seismic maps in R301.2, shown in Figures 1.2 and 1.4, respectively, the lateral load requirements can be determined for a given structure in a particular location. In some parts of the country the lateral loads for the design-level event (wind or seismic) are lower. In these parts of the country minimal bracing is required for residential-type structures. Other parts of the country may be in high wind or seismic regions, and some may be areas with both.

Who determines the wind and/or seismic potential for a given locale? Your local building officials have already determined your wind or seismic region (and other parameters for the area, including snow load, frost depth, termite danger, flood hazard, etc.). When your local jurisdiction adopted the IRC it was required to fill out Table R301.2(1) for Climatic and Geographic Design Criteria.

What is the lateral load path?

It is very important to understand this concept of a lateral load path as it helps make sense of the prescriptive requirements in the IRC. In short, the lateral load path is simply the path that the lateral or horizontal load takes as it passes through the structure, and its components and connections, on its way to the ground. See Figure 1.5 to understand the critical parts of the lateral load path. Vertical loads similarly move through structural components in their own load path to the ground.

A conceptual description of the lateral load path for wind is easier to understand than the load path for seismic (with inertially induced forces). For simplicity, the following figure focuses on an imposed load on a windward wall and its subsequent load path through the building.

FIGURE 1.5
CRITICAL PARTS OF THE LATERAL LOAD PATH

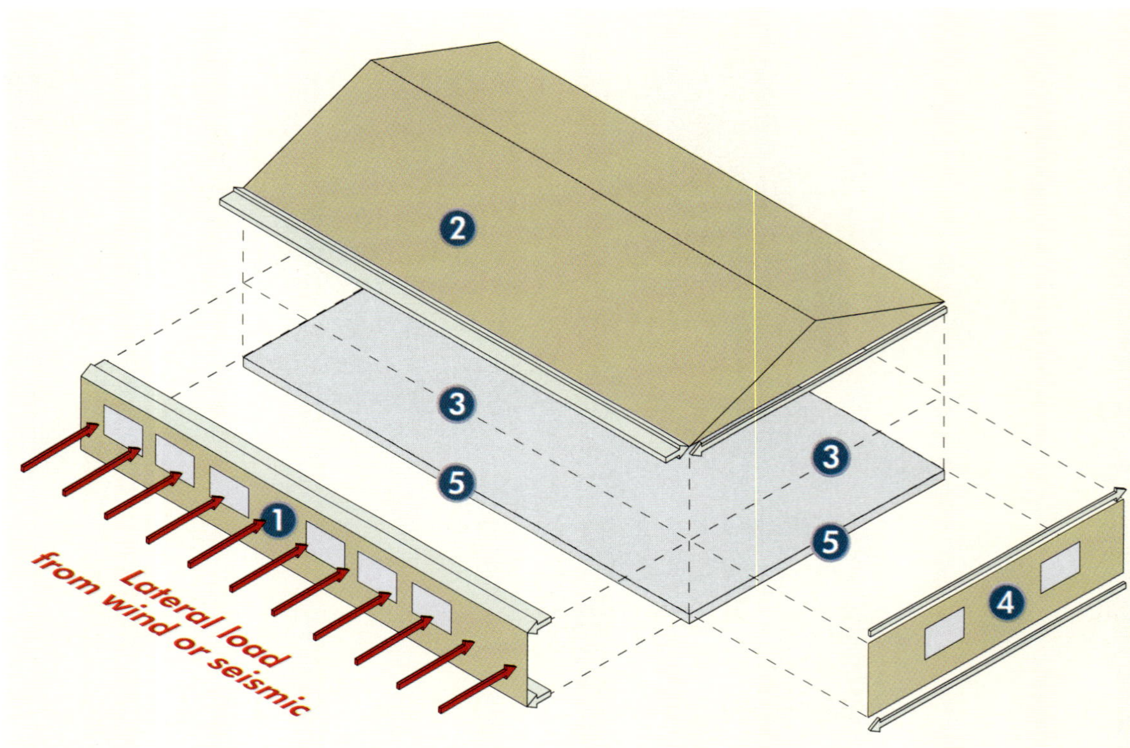

Critical parts of load path:

1. Receiving wall carries load to roof diaphragm at top and foundation at bottom
2. Roof or floor diaphragm carries load to bracing walls
3. Connections between roof/floor and wall
4. Bracing wall carries load from diaphragm to foundation
5. Connections between wall and foundation

As background, it may help to understand some commonly used terminology, such as diaphragms and shear walls. Section R202 of the IRC defines them as follows:

> ***Diaphragm.*** *A horizontal or nearly horizontal system acting to transmit lateral forces to the vertical resisting elements. When the term "diaphragm" is used, it includes horizontal bracing systems.*
>
> ***Shear Wall.*** *A general term for walls that are designed and constructed to resist racking from seismic and wind by use of masonry, concrete, cold-formed steel or wood framing in accordance with Chapter 6 of this code and the associated limitations in Section R301.2 of this code.*

In the IRC, diaphragms are simply roofs or floors.

Shear walls and braced wall segments do the same thing: they resist racking from lateral loads. In this book, "shear wall" will be used to refer to an engineered wall segment in accordance with the IBC, and "braced wall" will be used to refer to a wall segment constructed in accordance with the IRC. Further details about the differences between braced walls and shear walls are presented in Chapter 2.

In the vertical load path, structural members carrying gravity load (rafters, trusses, joists) all must bear on members (posts, cripple walls, beams, etc.) designed to carry that load. Each of these members must bear on others until the load is transferred into the ground through the foundation. The vertical load path is fairly easy to understand. Just as with our first set of building blocks, one block is essentially stacked on another. Due to the force of gravity, vertical loads are always present and it's relatively easy to identify the load path, as shown in Figure 1.6.

FIGURE 1.6
EXAMPLE OF VERTICAL LOAD PATH

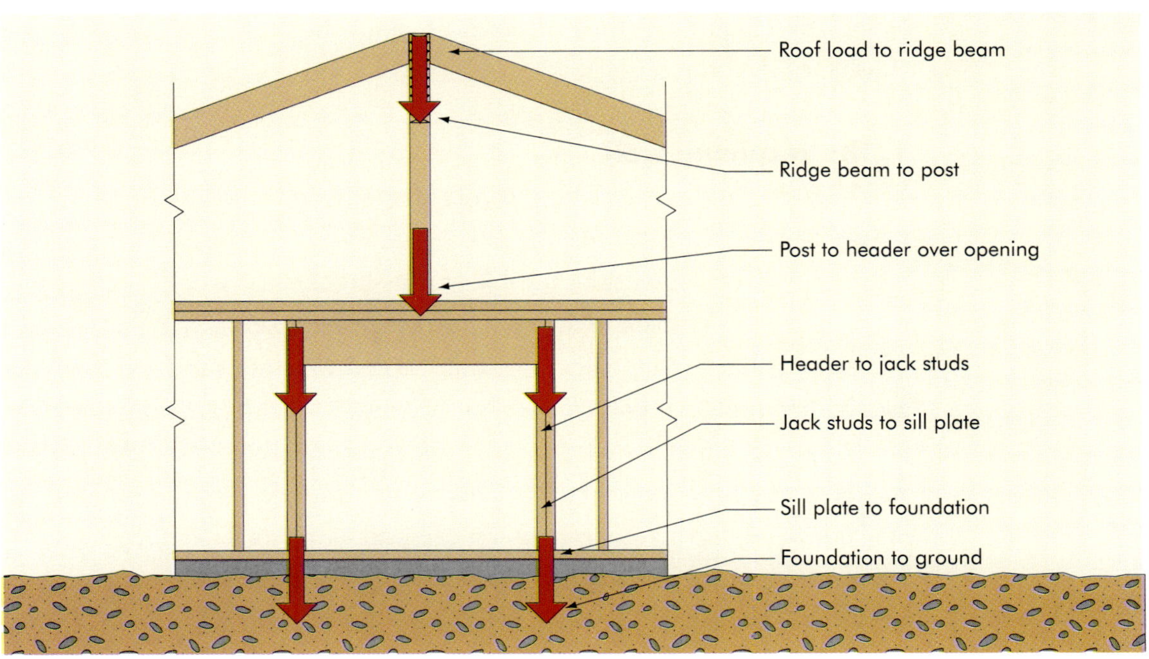

While the objective of the lateral load path (to get the applied loads into the ground) is exactly the same as the gravity load path, the actual path it takes is not always obvious. Another important difference is that with a lateral load path, connections are increasingly important. Unlike gravity in the vertical load path where many members are simply bearing on each other, there is nothing holding elements together in the lateral load path unless the builder puts something there to transfer the load, such as nails, straps, or framing anchors. Also, unlike the vertical loads, lateral loads are not constant, as the "design event" high winds and/or earthquakes are relatively uncommon occurrences. The problem is that when these lateral load events *do* occur, the lateral load path must be there with each element and connection properly sized and constructed.

Critical parts of the lateral load path

As shown in Figure 1.5, there are five critical parts in the load path for a simple rectangular structure. These are discussed in this chapter to explain how the wind and/or seismic loads are distributed through a simple rectangular structure and into the foundation. The critical parts of lateral load path are: 1) the receiving wall, 2) floor and roof diaphragm, 3) roof-to-wall/wall-to-wall connections, 4) wall bracing, and 5) wall bracing-to-foundation connections. The numbered areas in Figure 1.5 do not pertain to load path sequence, but are used to differentiate each part, and correlate the following disaster photos and discussion to the theoretical load path concept shown in Figure 1.5. Each part is critical. If any one element in the load path fails, the structure typically fails. A good analogy is a chain. Like a chain, the lateral load path is only as strong as the weakest link (receiving wall, diaphragm, bracing wall or the connections between these parts and the foundation). While the discussion focuses on a simple rectangular structure, as shown in Figure 1.5, the principles also apply for multi-story structures, and "T"- and "L"-shaped structures.

1. The receiving wall

The receiving wall is the wall perpendicular to the direction of the lateral load. These walls receive or catch the wind like the sail of a sailboat or, in the case of an earthquake, add seismic mass. The wall itself must be capable of resisting the loads, both suction and positive pressure, in order to transfer them to the next link in the load path. (Note that while the receiving wall is under positive pressure, the opposite and end walls are subjected to suction pressures. These suction pressures can pull off poorly attached wall cladding, siding, windows and doors.) This has additional importance for wind forces, because if the sidewall fails to withstand the wind suction pressure, it can lead to the failure of a structure's weatherproofing system. While this may not cause a structural collapse, it can lead to the total destruction of the contents of the house. Some examples of wind related failures at the receiving wall due to its inability to transfer the applied loads to the diaphragm above are shown in Figures 1.7 through 1.9.

FIGURE 1.7

Wall covering is an essential part of the first step of the load path for wind. The wall studs can be seen behind the failed wall covering system. The failure could have been due to various reasons. Approved wall coverings installed per code would have most likely distributed the wind load to the roof diaphragm. (Photo taken after Hurricane Katrina, inland, where wind speeds were within scope of IRC.)

FIGURE 1.8

Not all wall coverings are capable of resisting code-required wind pressures. (See Table R301.2(2).) This house was subjected to an 85 mph wind. The reason for failure could have been due to various issues, including improper installation or inadequate field-applied connections. (Photo taken after Hurricane Katrina, inland, where wind speeds were within scope of IRC.)

FIGURE 1.9

A partial failure of the wall covering system. Note also the house in the background, to the right, which had similar failure. (Photo taken after Hurricane Katrina, inland, where wind speeds were within scope of IRC.)

On the other hand, an earthquake acts on the entire structure and not just the receiving wall. The receiving walls, therefore, are not exposed to pressure, but add seismic mass that is spread throughout the structure. Failure of brick veneer has been observed where the brick veneer is inadequately tied to the rest of the structure. A major difference between failures in seismic and wind events is that wind-driven rain is not usually associated with seismic events.

Assuming the receiving walls themselves can withstand the force of the lateral load, the receiving walls transfer this load to the next component in the load path.

The receiving wall must be properly attached at the base (discussed in Section 5) and at the top (discussed below in Section 3) to adequately transfer the forces into the foundation (base attachment via anchor bolts or attachment per R602.3(1)) and into the roof diaphragm (attached per R602.3(1)). The connection between structural elements of the load path is just as important as the proper selection and detailing of the elements themselves. While discussed in greater detail below, proper inter-element connection cannot be over emphasized. Great attention to detail in these locations is very important in areas of high wind or moderate to high seismic force.

2. Floor and roof diaphragm

The floor and roof sheathing form the diaphragms which transfer load from the receiving wall to the bracing walls. Floor and roof *diaphragm* failures are rare. When failures do occur, especially in roof systems, they are often due to inadequate attachment of the roof sheathing, particularly in areas vulnerable to high wind pressure, such as gable ends and roof overhangs, as shown in Figure 1.10.

FIGURE 1.10

The sheathing was not fastened properly to the roof framing. The loss of sheathing compromises the diaphragm. (Photo taken after a tornado in Southwestern Missouri, May 2003.)

3. Roof-to-wall/wall-to-wall connections

Connecting the roof-to-walls and walls-to-walls (in the case of multi-stories) is critical to the transfer of loads. A common failure in the load path is the roof-to-wall or wall-to-wall connections. Figures 1.11 through 1.14 demonstrate this failure mode. As the walls support the roof for both vertical and lateral loads, such a failure can lead to total building collapse. For roof-to-wall connections it is important to realize that the roof diaphragm effectively begins and ends at the wall lines and the connection between the roof and wall at this point is critical to an effective load path. Connections of walls-to-floor and floor-to-walls are made in accordance with R602.3(1). It is important to realize that the attachment requirements of R602.3(1) are minimum connections. If the minimum connection is going to be used, it is essential that every fastener be properly placed.

FIGURE 1.11

Notice that the walls are still intact, just no longer in place. This was a panelized building and the wall panels were not attached sufficiently to resist the wind loads. (Photo taken after Hurricane Andrew.)

FIGURE 1.12

This is an example of a garage door failure (in background) that led to pressurization of the garage wall. Insufficient attachment at the top and bottom caused the wall to blow out. This portion of the end wall is no longer available to provide bracing for the rest of the structure. (Photo taken after Hurricane Katrina.)

FIGURE 1.13

This figure shows insufficient attachment of the end walls of the structure. It also shows failures in the wall covering products used to weatherproof the structure. (Photo taken after a tornado in Evansville, Indiana, 2005.)

FIGURE 1.14

This figure shows a case where the structure failed due to insufficient attachment of numerous structural elements, including receiving wall-to-roof, roof-to-wall, and wall-to-wall. The entire roof of this structure was lifted off in one piece and struck the house across the street, immediately behind the cameraman. (Photo taken after Hurricane Iniki.)

For multi-story structures in high wind areas, builders may be accustomed to using strap-type anchors between floors. These greatly increase the structure's ability to stay together. While often used for uplift only, the straps also provide augmentation to the lateral load path, if for no other reason than to prevent uplift-type failures from damaging the lateral load path.

4. Wall bracing

From a lateral load perspective, the walls support the roof and floor diaphragms through the use of bracing panels. The type, amount, and frequency of bracing panels are, of course, dependent on the level of the lateral load. Stronger and more frequent bracing walls are required in areas of high wind and seismic activity. Racking occurs when the wall is not strong enough to resist lateral in-plane loads – loads that are parallel to the direction of the wall. Racking occurs when a rectangular wall shifts to a parallelogram shape, where the top and the bottom of the wall are parallel but the sides are no longer vertical. The purpose of bracing is to prevent racking. Examples of racking in various degrees are shown in Figure 1.15.

FIGURE 1.15

FAILURES IN WALL BRACING AS INDICATED BY WALL RACKING

(File photos, both wind and seismic events.)

5. Wall-to-foundation connections

The receiving wall must be attached to the foundation in order to resist the loads imposed, just as the bracing walls must be attached to the foundation in order to resist the loads imposed. These connections are critical. Some examples of inadequate connection between the walls and foundation are demonstrated in Figures 1.16 and 1.17.

FIGURE 1.16

Insufficient (or non-existant) wall anchorage of the walls to the foundation caused the whole structure to be pushed off the foundation in this wind event. Additional connection failures can be seen in the background. (Photo taken after a tornado in Southwestern Missouri, May 2003.)

FIGURE 1.17

This picture is a close-up of insufficient foundation anchorage. Few bolts were used and washers were not used under nuts. Direction of bends indicated a shear failure of the bottom plate in a direction up and to the right hand corner of the picture. (Photo taken after a tornado in Southwestern Missouri, May 2003.)

The solution

While the previous photos suggest that damage is inevitable in lateral load events, that is not always the case. In every example shown, the damage to the building could have been prevented if the structure had been designed or built in accordance with the prevailing building code. In almost every example, there were other, similar structures in very close proximity that sustained little or no damage.

Each chapter of this book will explain in detail how to use the IRC to provide adequate wall bracing so that the house can be built to resist lateral loads. Figures 1.18 and 1.19 show only two examples of how properly built homes can survive even the most severe wind or seismic events.

FIGURE 1.18

After the 1964 Good Friday earthquake in Alaska, a house built with continuously sheathed plywood retained its box-like structure, even after a significant portion of the ground below it shifted more than 20 feet. Note that the walls are strong enough to cantilever the structure over the large void. The house to the left appears to have performed well without any noticeable damage. (Photo taken after the Great Alaska Earthquake, 1964.)

FIGURE 1.19

After Hurricane Andrew in Florida (1992), a waterfront house stands in near perfect condition, even after being battered by significant winds and water. Such cases of survival demonstrate that a proper load path works. (Photo taken after Hurricane Andrew.)

CHAPTER 2

History of Wall Bracing and How it Works to Resist Forces

As discussed in the preceding chapter, wind and seismic events subject structures to loads acting parallel to the ground (lateral loads). Regardless of how the structure is designed, whether through the use of prescriptive braced walls or engineered shear walls, elements must be built into the structure to resist these lateral loads or there is a potential for structural failure. Both braced walls and shear walls serve the same purpose in the load path discussed in the previous chapter.

Braced Walls (Prescribed) **vs. Shear Walls** (Engineered)

Bracing provisions in the model codes are prescribed for "conventional" construction. From the user's perspective, there is no engineering in conventional construction since the "solution" is prescribed. For structures or portions of them that do not meet conventional construction parameters, however, engineered design per the International Building Code (IBC) is required. The engineered versions of braced walls are called shear walls. Both provide racking resistance to lateral loads but each is part of distinctly different model-code provisions.

Prescriptive vs. Engineered

Prescriptive Construction

Limitations:

- 3 stories max.
- Wind < 110 mph (< 100 mph in hurricane-prone regions)
- SDC A – D_2
- Many others (see IRC Chapter 3)

Typically uses braced wall panels ***without*** hold downs to prevent walls from racking.

Engineered Design

Applications:

- Any size/shape within IBC limits
- Wind – No limit
- Seismic – No limit
- Governed by engineers' calculations

Typically uses shear walls ***with*** pre-engineered hold downs (other than anchor bolts) to prevent walls from racking.

What's the difference between a braced wall panel and shear wall?

Braced wall panels, the primary focus of this book, are part of conventional prescriptive construction. In the conventional construction provisions of the IRC, wall bracing is prescribed – the user selects the bracing amount required directly from a table in the IRC – and few, if any, calculations are required.

Shear walls are designed by an engineer and have specific design values depending on fastener spacing, fastener size, sheathing thickness, and framing species. They usually require pre-engineered hold downs to resist overturning of wall elements. The shear wall must resist the loads that are calculated by the engineer's analysis.

What is bracing and how does it work?

The framing elements of a typical wall are shown in Figure 2.1. When subjected to lateral loads the stud wall alone has very little racking resistance with each of the joints acting as a hinge (shown as a small circle in Figure 2.1). It takes very little lateral load to rack the wall (turn it into a parallelogram).

FIGURE 2.1

BARE STUD WALL HAS NO LATERAL LOAD RESISTING CAPACITY

In conventional construction there are two ways commonly used to provide a wall with lateral load (racking) resistance. The first way is to use a diagonal 1x4 let-in brace between 45 and 60 degrees to the horizontal as seen in Figure 2.2, and the second, through the attachment of relatively rigid rectangular panels as seen in Figure 2.4. (There is a third method permitted, the use of diagonal wall boards, that is seldom used today. Due to its limited use, it is not covered in subsequent discussions.)

Let-in bracing

Let-in bracing, known in the code as Method 1 bracing and shown in Figure 2.2, is the original type of bracing that was used before panel-type sheathing products were developed.

The effectiveness of let-in bracing is dependent to a great extent on the craftsmanship of the builder cutting out the notches to let-in the 1x4, and by the nails that attach it to the top and bottom plates. Since there are only two nails at each point where the 1x4 crosses the framing, the strength of this bracing method is limited.

To further complicate understanding the effectiveness of let-in bracing, it is always used in conjunction with other structural sheathing products. Originally, horizontal board siding was used. Today, gypsum wall board sheathing is most common.

FIGURE 2.2

STUD WALL WITH LET-IN BRACE

The diagonal brace and two nails into the top and bottom plates and each stud it crosses prevent the wall from racking.

A number of manufacturers have developed proprietary metal straps designed to function like a 1x4 brace. These either require no modification of the studs or just a shallow saw kerf in the studs to facilitate their installation. Generally speaking, the thin flat straps may have to be used in an "X" pattern while the straps with an "L" or "T" cross section may only need to be used in a single direction, like let-in bracing. These proprietary products are recognized for use in code-conforming construction via ICC-ES or Legacy reports. Copies of these reports, which include use restrictions, can be downloaded free of charge from the International Code Council – Evaluation Services web site at icc-es.org. It is important to read the manufacturer's installation recommendations carefully because some of these strap-type anchors are approved for use as bracing and some are meant for temporary bracing during construction only.

Panel-type bracing (and portland cement lath and plaster)

As mentioned on the previous page, let-in bracing and the strap-type proprietary products have only limited structural capacity. As such, the IRC limits their use to areas of low wind and seismic load. Of considerably more capacity, and consequently of greater use for bracing, are panel-type products.

All panel-type products work in the same manner, while varying greatly in capacity from wood structural panel products (Method 3 bracing) at the upper end of capacity to gypsum wallboard (Method 5 bracing) at the lower end.

- The panel-type products are manufactured in a rectangular shape and, through their physical properties, resist being forced out of this shape. See Figure 2.3.
- The panel products have fasteners spaced around their entire perimeter with spacing usually ranging from 6 to 3 inches o.c. depending on the bracing method.
- When these panels are attached to a stud wall, such as the one shown in Figure 2.1, they provide racking resistance, as shown in Figure 2.4.
- Portland cement lath and plaster is included with panel siding because, although "manufactured" in place, when completed, it performs in a similar manner as panels.

FIGURE 2.3

Rectangular panel products (and portland cement lath and plaster) resist being forced out of shape by their physical properties and nail attachment to studs.

FIGURE 2.4

STUD WALL WITH PANEL BRACING

Perimeter nailing prevents the wall from racking.

The ability of a panel type to resist loads is dependent on the physical properties of the panel, primarily its strength and rigidity. The ability of a panel type to impart this resistance onto the wall is dependent primarily on the attachment of the panel to the wall studs around the perimeter of the panel. The attachment capacity is dependent on a number of variables, including size of nails (or other fastener), number of nails, placement of nails, framing lumber species, and, again, the physical properties of the panel, such as fastener bearing strength, tear resistance, and fastener pull through resistance.

Variations in bracing panel physical properties and prescribed attachments account for the different use requirements for the various panel types. These differences are discussed in more detail in Chapter 5.

History of wall bracing

Wall bracing requirements are not new to the codes. In fact, several of the current bracing methods, including diagonal boards and plaster, reflect conventional practices that were more common 50 years ago.

Wall bracing requirements date back to the first Uniform Building Code (UBC), 1927 edition. The 1927 UBC required:

> ***Section 2205*** (4th paragraph). *All exterior walls and partitions shall be thoroughly and effectively angle braced.*

Similar requirements were in Section 2507(e).

In the 1920s, panel sheathing products were not used. Thus, the means of bracing was presumably let-in bracing, or some other type of angle bracing.

The UBC bracing provisions remained unchanged until 1952, which was also the year that vertical and horizontal diaphragm tables appeared. The text of the 1952 UBC is virtually identical to the 1927 UBC, however, plywood panels were added:

> ***2521(f) –*** *...All exterior walls and partitions shall be thoroughly and effectively angle braced* ***or sheathed with approved panels adequately nailed along all edges.***

These provisions were unchanged until 1970. The 1970 UBC added several pages to describe conventional construction provisions. These sections included expanded descriptions of bracing types and prescriptive provisions, such as the distance between bracing panels. The text of the 1970 UBC bracing provisions in Section 2518 was:

> ***5. Bracing.*** *All exterior walls and main cross stud partitions shall be effectively and thoroughly braced at each end, or as near thereto as possible, and at least every 25 feet of length by one of the following methods:*
>
> A. *Nominal 1-inch by 4-inch continuous diagonal braces let into studs...* (nearly identical to today's Method 1)
>
> B. *Wood boards of 5/8-inch net minimum thickness applied diagonally on studs...* (nearly identical to today's Method 2)
>
> C. *Plywood sheathing...* (nearly identical to today's Method 3)
>
> D. *Fiberboard sheathing...* (nearly identical to today's Method 4)
>
> E. *Gypsum sheathing...* (nearly identical to today's Method 5)
>
> F. *Particleboard sheathing...* (nearly identical to today's Method 6)
>
> *For methods B, C, D, E and F, the braced panel must be at least 48 inches in width...*

In the 1994 UBC, a 32-inch alternate braced wall panel was added that prescribed sill plate anchorage, foundation details, and hold-down devices. This 32-inch alternate braced wall panel with hold downs is still in today's codes. The alternative braced wall panel was essentially a prescriptive shear wall that was "designed" to be equivalent to a Method 3 wall segment and thus, is a suitable substitution for all bracing methods.

Between the 1970 and the 1997 UBC, *"effectively and thoroughly braced"* was more clearly defined.

Other model codes, such as the Council of American Building Officials (CABO) International One- and Two-Family Dwelling Code (IO&TFDC), the Southern Building Code Congress International (SBCCI) Standard Building Code, and the Building Officials and Code Administrators International (BOCA) National Building Code, also had similar bracing provisions and evolution.

During the development of the 2000 IRC, the IRC Code Development Committee selected and combined code provisions from the existing model codes to form the International Residential Code.

Some of the new notable wall bracing provisions in the 2000 IRC were:

- Wall bracing was required to meet a minimum percentage of the braced wall line. The amounts depend on the load (wind speed or seismic design category (SDC)) and how many stories are above the braced wall line.
- A continuous wood structural panel bracing method was added, permitting wall bracing elements as narrow as 24 inches to be used to meet the percent bracing requirements.

In the 2003 IRC, braced wall line spacing limits were added. Also, the continuous wood structural panel sheathing option (based on the perforated shear wall method defined in the IBC) was specified for use on all exterior and interior braced wall lines.

In the 2006 IRC, the 32-inch-wide alternates were changed to permit wall bracing elements as narrow as 28 inches for certain cases. A portal frame with hold downs was added which permitted bracing elements as narrow as 16 inches for certain cases. Also, the portal frame was permitted to be used without hold downs for bracing next to garage doors (in Seismic Design Categories A–C, for use with up to one story above) when used in conjunction with continuously sheathed wood structural panel walls (Table R602.10.3, Footnote c). Each of the wall bracing methods in the 2006 IRC is described in detail in Chapter 5.

In short, model codes have required wall bracing for at least 70 years. The changes to wall bracing provisions have accelerated in recent years.

TABLE 2.1

Summary of wall bracing history in model codes

Year	Bracing Provisions
1920s	"…thoroughly and effectively angle braced."
1950s	"…thoroughly and effectively angle braced or sheathed…"
1970s	"…effectively and thoroughly braced at each end and … every 25 feet … by one of the following methods…"
	"…braced panel must be at least 48 inches in width…"
1994	ADD: 32-inch alternate braced panel = 48-inch braced panel
2000	ADD: Minimum bracing percentages (based on load and story location)
	ADD: Continuous sheathed method
2003	ADD: Braced wall line maximum spacing
2006	ADD: Expanded alternate braced panels
	ADD: Portal frames

Why are bracing provisions changing?

The simple answer is that we don't build houses the way we used to!

In the 1960s, when more prescriptive bracing provisions began appearing in model building codes, houses were much smaller, seldom over two stories, had a greater number of interior walls and fewer windows. See Figure 2.5.

FIGURE 2.5

Typical single-family residence built in the 1960s.

Fast-forward to today and we see that houses are commonly up to four times larger than those built in the 1960s, have multiple stories, two-story high entrances, great rooms, window walls, huge three-car garages, complex roof lines and dramatic stairways that put big holes in the floor diaphragms. Any one of these features can have a negative impact on the structural performance of a building if not accounted for in the structural frame. Figure 2.6 is an example of the type of home being built today.

FIGURE 2.6

A residence more typical of the type being built today.

Another factor that must be taken into consideration is the greater prevalence today of nonstructural sheathing products. In the 1960s, the building industry had little concern about wall bracing because the majority of sheathing products available for use at that time were structural to one degree or another. Today, there are a number of nonstructural foam sheathing products that are popular for reasons of energy conservation. Many builders are compelled by market pressures to balance the structural requirements of the building with the requirements for energy savings. As a result, builders often inadvertently minimize the structural sheathing in an effort to maximize the insulating materials used. This trend, and other changes in modern residential design, are driving the seemingly endless complications of the prescriptive requirements.

For these reasons, along with high loads, many homes constructed in certain regions are engineered. For example, in Tacoma, Washington, 90 percent of all houses constructed have an engineered load path. This is also true for parts of Florida, California, Oregon, and other states.

Many of the changes to wall bracing in the last 15 years have been attempts to give some credit to narrower segments, since multiple 4-foot segments are not always available in today's home designs.

One thing that has not changed, however, is the necessity to insure that the new housing stock is safe. The imperative for safety fueled the evolution of building codes in the 1950s and 1960s. This evolution will continue as tastes, family structures and energy availability change in the future. Similarly, changes to the bracing provisions in the IRC will continue to parallel the evolving housing stock. Today's houses are very different from those built 30 years ago, and the houses of the future will likely differ significantly from today's home designs.

CHAPTER 3

Bracing Terminology and the International Residential Code

So far we have covered why we need bracing and how it works to resist lateral loads. With this information in mind, we are going to start relating this knowledge specifically to the bracing provisions of the IRC.

Bracing terminology

The first step is to get the terminology straight. When describing wall bracing, the IRC uses the terms **braced wall panel, braced wall panel spacing, braced wall line, and braced wall line spacing.** The user must understand what each of these terms means in order to follow the requirements in the IRC. Figure 3.1 shows how these terms relate to an actual structure.

FIGURE 3.1

BRACED WALL PANELS, BRACED WALL PANEL SPACING, BRACED WALL LINES AND BRACED WALL LINE SPACING

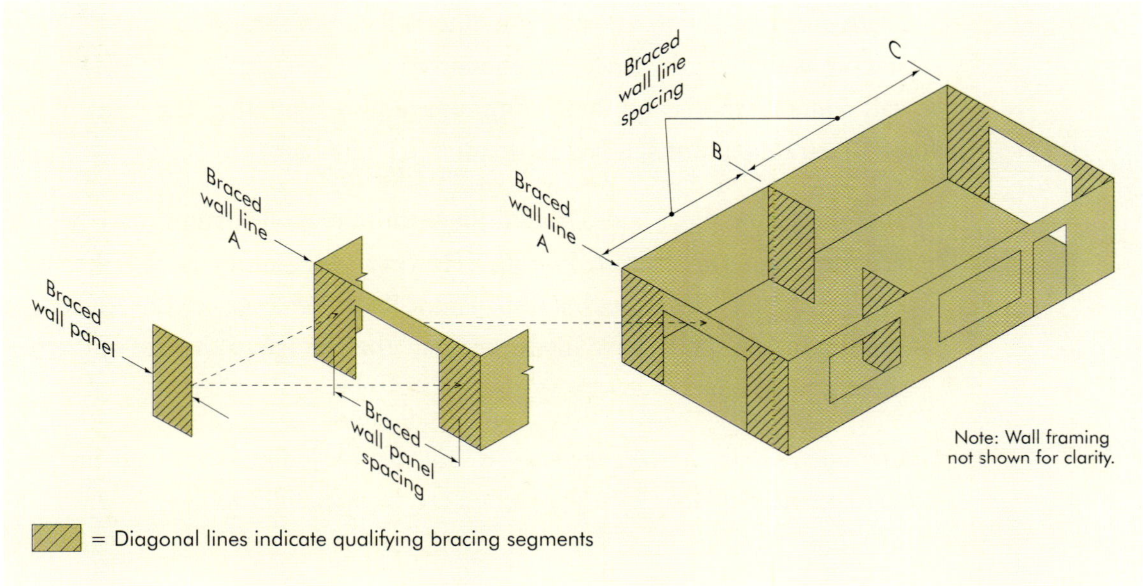

Braced wall panel

From Section R202 of the IRC:

> *"**BRACED WALL PANEL.** A section of a braced wall line constructed in accordance with Section R602.10 for wood framing or Section R603.7 or R301.1.1 for cold-formed steel framing, which extend the full height of the wall."*

Put in simpler terms, **braced wall panel** describes a code-qualified bracing segment. The name is probably derived from the fact that most of the recognized methods of bracing use panel-type products. Even let-in bracing is specified as a bracing panel even though let-in bracing is not a panel product. "Panel" is somewhat of a misnomer since it actually describes a wall "section," "segment," or "unit." As such, the terms braced wall section, braced wall segment, or bracing unit are often used interchangeably.

Each braced wall panel must extend the full height of the wall – from bottom plate to the top of the double top plates. The dimension of the braced wall panel measured in the direction parallel to the wall line is the *length* of the panel, from a starting point on the left to an ending point on the right of that panel. In other words a 4x8 oriented strand board bracing panel placed with the 8-foot dimension in the up-and-down direction is 4 feet long.

There are eight different bracing methods listed in the IRC and each has its own minimum bracing panel length. Table 3.1 contains a brief description of the IRC-permitted braced wall panel methods and their minimum required lengths. The bracing methods permitted are illustrated and discussed in detail in Chapter 5.

Note that some prior codes mixed the terminology of "length" and "width." Even the 2006 IRC uses the term "sheathed width." However, for simplicity, this discussion will use the term "length" to describe this dimension. Recent changes to the IRC at the Final Action Hearing in May 2007 have made "official" the change to the use of the term "length" in the bracing provision.

Alternatives to the bracing methods in Table 3.1, including several narrow, prefabricated solutions, have become available. Visit www.icc-es.org for more information.

TABLE 3.1

IRC bracing methods and minimum braced panel length[a]
(See referenced sections for detailed requirements.)

IRC Section	Bracing Method	Construction Description	Minimum Length of Braced Wall Panel
R602.10.3	1	Nominal 1x4 let-in bracing	55"-96"[b]
	2	Wood boards of 5/8" net thickness applied diagonally	48"
	3	Wood structural panel sheathing	48"
	4	Fiberboard sheathing	48"
	5	Gypsum sheathing	96"[c]
	6	Particleboard sheathing	48"
	7	Plaster	48"
	8	Hardboard panel siding	48"
R602.10.5	Continuous Sheathing	Wood structural panel sheathing	24", 16"[d]
R602.10.6.1	Alternate	Wood structural panel sheathing with hold downs	28"
R602.10.6.2	Portal Frame	Wood structural panel sheathing portal frame with hold downs	16"

a. Braced panel length is panel measurement parallel to the wall line.

b. Length of diagonal bracing depends on angle installed and the height of the wall. IRC requires installation of let-brace between 45 and 60 degrees from the horizontal.

c. 96" sheathed one side and 48" sheathed two sides.

d. Wood structural panel sheathing portal frame per Table R602.10.5, footnote c.

Braced wall panel spacing

The IRC, in Table R602.10.1, requires bracing panels to be spaced not farther than 25 feet o.c., as shown in Figure 3.2. The 25-foot spacing is measured from the centerline of the **bracing panel** to the centerline of the next closest bracing panel in the same wall line. See Figures 3.2 and 3.3.

Note: Measuring from the closest 4-foot bracing portion of a wider bracing segment as shown in Figure 3.3 is an interpretation of the code as it is currently written. As the intent of the code is to limit the unbraced portions of the wall to, in most cases, 21 feet – the clear distance between two 4-foot sections – the solution provided by Figure 3.3 is reasonable.

FIGURE 3.2

BRACED WALL PANEL SPACING

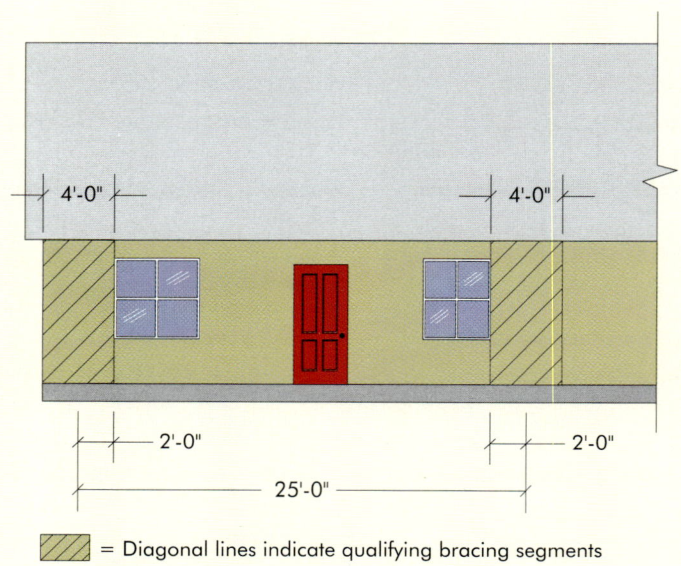

FIGURE 3.3

BRACED WALL PANEL SPACING WHEN BRACED WALL SECTIONS ARE MADE UP OF MORE THAN ONE PANEL

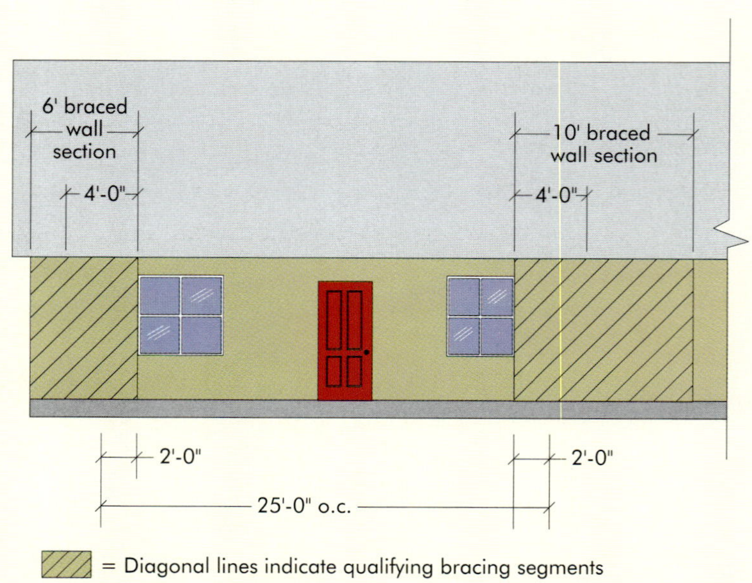

Braced wall line

From Section R202 of the IRC:

> ***"Braced Wall Line.*** *A series of braced wall panels in a single story constructed in accordance with Section R602.10 for wood framing or Section R603.7 or R301.1.1 for cold-formed steel framing to resist racking from seismic and wind forces."*

Figure 3.1 shows three separate braced wall lines, A, B and C. For a simple rectangular building, as shown in Figure 3.1, the exterior walls perpendicular to braced walls lines A, B and C are also braced wall lines. Each wall line is made up of a number of braced wall panels as shown in Figure 3.1. Unlike the figure, each braced wall line does not have to be "in a line." The code permits offsets of up to 4 feet (Section R602.10.1). This is in recognition of the assumption that if the wall lines are offset by 4 feet or less, they act in combination to resist lateral load. If they are farther apart than 4 feet, they must be considered as separate braced wall lines. Offsets in braced wall lines are further discussed in Chapters 7 and 10.

Braced wall lines are not always exterior walls, as braced wall line B in Figure 3.1 illustrates. If the house is big enough, due to braced wall line spacing requirements, the code requires *interior* braced wall lines to supplement the exterior wall lines to maintain the structure's ability to resist lateral loads. Each of these interior braced wall lines must be treated just like the exterior wall lines in terms of percent of bracing, panel location, and offsets. The only exception is in the braced wall line foundation requirements. In lower seismic regions, the foundation requirements are not as stringent for interior braced wall lines. These and other attachment requirements are covered in Chapter 8.

Braced wall line spacing

As shown in Figure 3.1, braced wall line spacing is the spacing between parallel braced wall lines. If the maximum spacing provided by the code is exceeded, interior braced wall lines must be provided as described above. The braced wall line spacing requirements in the IRC are given below:

For all wind areas covered by the IRC (0 to less than 110 mph or 100 mph in hurricane-prone regions) and Seismic Design Categories A through C, Section R602.10.1.1 provides requirements on the maximum spacing between **braced wall lines.** This spacing is limited to a maximum of 35 feet with an exception that permits an increase of up to 50 feet if both of two criteria are met:

> *"1. The wall bracing installed equals or exceeds the amount of bracing required by Table R602.10.1 multiplied by a factor equal to the braced wall line spacing divided by 35 feet, and*
>
> *2. The length-to-width ratio of the floor or roof diaphragm does not exceed 3:1."*

In Section R602.10.1.1, Exception 1 quoted on the previous page, the wall bracing that must be increased is the wall bracing perpendicular to (at right angles to) the 35-plus foot dimension, as shown in Figure 3.4. This is logical because as the distance between the bracing elements increases, the load on the diaphragm increases. The load in the walls supporting that diaphragm – the walls perpendicular to the 35-plus foot dimension – also increases.

FIGURE 3.4

EXAMPLE FOR CALCULATING THE INCREASE IN BRACED WALL LINE PERCENTAGE FOR WALL LINE SPACING GREATER THAN 35 FT PER R602.10.1.1

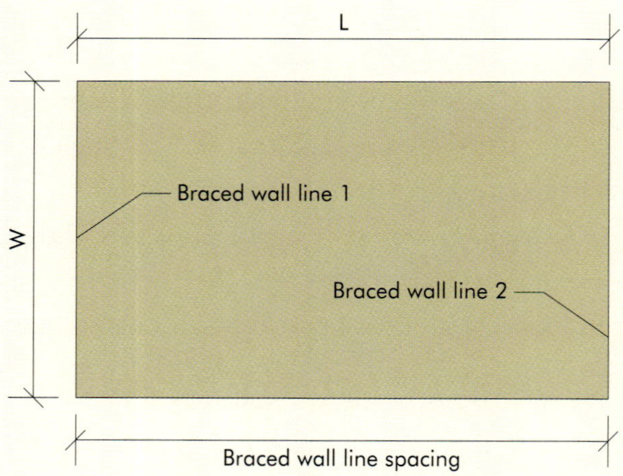

1) For braced wall line spacing of greater than 35 ft, the amount of bracing required for braced wall lines 1 and 2 per Table R602.10.1 is braced wall spacing in feet divided by 35.

   For example:
   For a 45-ft braced wall line spacing (L = 45 ft), the required amount of bracing for braced wall lines 1 and 2 per Table R602.10.1 would be increased by: 45/35 = 1.29

2) L:W ratio must not exceed 3:1

   For example:
   For L = 45 ft, W must be 15 ft or greater.

Braced wall line spacing requirements apply to all braced wall lines, interior and exterior.

In high seismic areas (Seismic Design Categories D_0, D_1, and D_2) the maximum distance between braced wall lines is reduced from 35 feet to 25 feet per Section R602.10.11.1. There is an exception to this minimum length as well:

> ***"Exception:*** *In one- and two-story buildings, spacing between two adjacent braced wall lines shall not exceed 35 feet (10 363 mm)(sic) o.c. in order to accommodate one single room not exceeding 900 square feet (84 m²) in each dwelling unit. Spacing between all other braced wall lines shall not exceed 25 feet (7620 mm)."*

The above exception permits for each dwelling unit a one-time increase for ***one*** braced wall line spacing to 35 feet to accommodate a 900-square-foot room. This one-time increase is applicable to any part of the contiguous house including the attached garage. For example, if this exception is taken for a three-car garage, it cannot also be taken for a large great room.

CHAPTER 4

Loads and Limits of the International Residential Code

Because the International Residential Code (IRC) is a prescriptive code, certain limitations are placed on its use. In terms of geographic locations, these limitations include areas with Seismic Design Categories (SDC) A–D_2, and wind regions less than 110 miles per hour (100 miles per hour in hurricane-prone regions).

In addition to the above limitations, Chapter 3 of the IRC has a number of geometric limitations on the size, shape and architectural features of the structure. As the IRC is prescriptive in nature, it is not possible to cover every combination of non-square walls, cantilevers, large wall line offsets, three-sided structures, and multi-story residences. The Chapter 3 limitations have been developed so that the prescriptive provisions of the IRC will result in safe structures. Structures or elements of structures that are outside of the limitations of Chapter 3 must be designed in accordance with the International Building Code (IBC) structural provisions or other referenced documents.

In this chapter, we'll discuss wall bracing considerations as they relate to limitations of the IRC:

1. The scope of the IRC and what to do when details of the structure go beyond the IRC
   a. What to do and where to go when the building is outside the scope of the IRC
   b. IRC requirements for alternate methods
   c. Mixing engineered design with prescriptive construction
   d. Other referenced alternate provisions

2. Construction systems
   a. Balloon and platform framing

3. Story height
   a. Story height limits

4. Design criteria – what are the local design requirements?
   a. Determining the requirements for a particular site

5. Wind loads
   a. Limitations
   b. Component and cladding wind pressure requirements
   c. Wind Exposure Categories

6. Seismic loads
    a. Seismic exemptions
    b. IRC seismic basics
    c. Seismic limits: – Seismic Design Categories A–D_2
        a) dead load
        b) building irregularities
        c) stone and masonry veneer
        d) number of stories
    d. Additional seismic limitations – Seismic Design Category C
    e. Additional seismic limitations – Seismic Design Categories D_0, D_1 and D_2

7. Snow loads
    a. Ground snow load limit

The following sections provide greater detail and discussion of the outlined topics.

1) The scope of the IRC and what to do when details of the structure go beyond the IRC

For the residences that are outside of the geographic or geometric scope of the IRC, the designer can use the structural provisions of the IBC. Not only may the IBC be used to completely design residential structures, it may be used to design any *part or portion* of the structure that is outside of the scope of the IRC. This means that the IRC and IBC can be used in combination to design a structure. This is permitted by the following sections:

> ***R104.11 Alternative materials, design and methods of construction and equipment.*** *The provisions of this code are not intended to prevent the installation of any material or to prohibit any design or method of construction not specifically prescribed by this code, provided that any such alternative has been approved. An alternative material, design or method of construction shall be approved where the building official finds that the proposed design is satisfactory and complies with the intent of the provisions of this code, and that the material, method or work offered is, for the purpose intended, at least the equivalent of that prescribed in this code. Compliance with the specific performance-based provisions of the International Codes in lieu of specific requirements of this code shall also be permitted as an alternate.*
>
> *Note: Section R104.11 is also the enabling language typically used to evaluate alternative products via evaluation reports such, as those found at www.icc-es.org.*

> ***R301.1.3 Engineering design.*** *When a building of otherwise conventional construction contains structural elements exceeding the limits of Section R301 or otherwise not conforming to this code, these elements shall be designed in accordance with accepted engineering practice. The extent of such design need only demonstrate compliance of non-conventional elements with other applicable provisions and shall be compatible with the performance of the conventional framed system. Engineering design in accordance with The International Building Code is permitted in all buildings and structures, and parts thereof, included in the scope of this code.*
>
> ***R602.10.10 Design of structural elements.*** *When a building, or portion thereof, does not comply with one or more of the bracing requirements in this section, those portions shall be designed and constructed in accordance with accepted engineering practice.*

In addition to using the IBC to design parts or portions of the structure, Section R301.1.1 – Alternative provisions of the IRC, permits the use of the following for wood construction:

- American Forest and Paper Association (AF&PA) *Wood Frame Construction Manual* (WFCM) as an alternative to the provisions in the IRC.

In IRC Section R301.2.1.1 – Design Criteria, the code requires, for regions *outside* of the basic winds speeds covered by the IRC, designs based on one or more of the following:

- American Forest and Paper Association (AF&PA) *Wood Frame Construction Manual* (WFCM),
- Southern Building Code Congress International Standard for Hurricane Resistant Residential Construction (SSTD-10), or
- American Society of Civil Engineers Minimum Design Loads for Buildings and Other Structures (ASCE-7).

2) Construction systems – balloon and platform framing

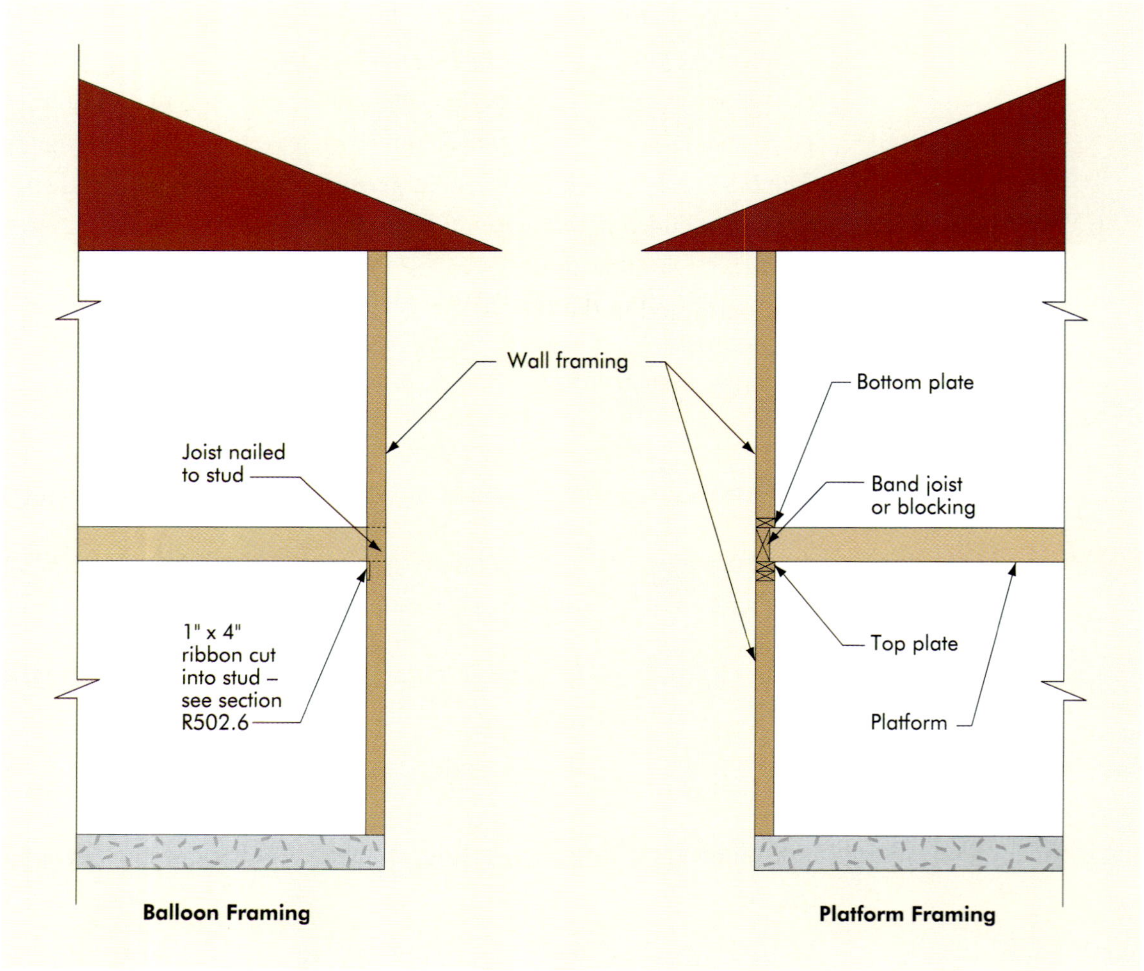

FIGURE 4.1
BALLOON AND PLATFORM FRAMING

The requirements of the IRC for light-frame construction apply to both balloon- and platform-framed buildings (R301.1.2). While this may seem to be only an issue with multi-story residences, some jurisdictions within the U.S. require balloon framing at gable ends of single story residences. These jurisdictions are normally located in high wind regions of the U.S. and the balloon framing is required to prevent a "hinge" from forming between the end wall and the gable wall framing above when the wind blows directly on the wall. This has historically been a problem in high wind/hurricane-prone regions.

The studs in balloon-framed walls extend from the lower story sill plate to the upper story top plate. Upper floors are typically attached to the walls with a ledger attached to the sides of the wall studs. For platform-framed walls, the studs extend from the sill plate to the top plate for each story, and upper floors are built above lower story walls. Figure 4.1 illustrates the difference between balloon and platform frame construction.

3) Story height limits per R301.3

IRC Section R202 gives the definition for story height as follows:

> ***Story height.*** *"The vertical distance from top to top of two successive tiers of beams or finished floor surfaces; and, for the topmost story, from the top of the floor finish to the top of the ceiling joists or, where there is not a ceiling, to the top of the roof rafters."*

Figure 4.2 shows story height for light-framed wood and cold-formed steel construction. Figure 4.3 shows a similar figure for masonry construction. Table 4.1 provides a summary of wall heights permitted in the IRC.

FIGURE 4.2
STORY HEIGHT MEASUREMENT WOOD AND COLD-FORMED STEEL FRAMING

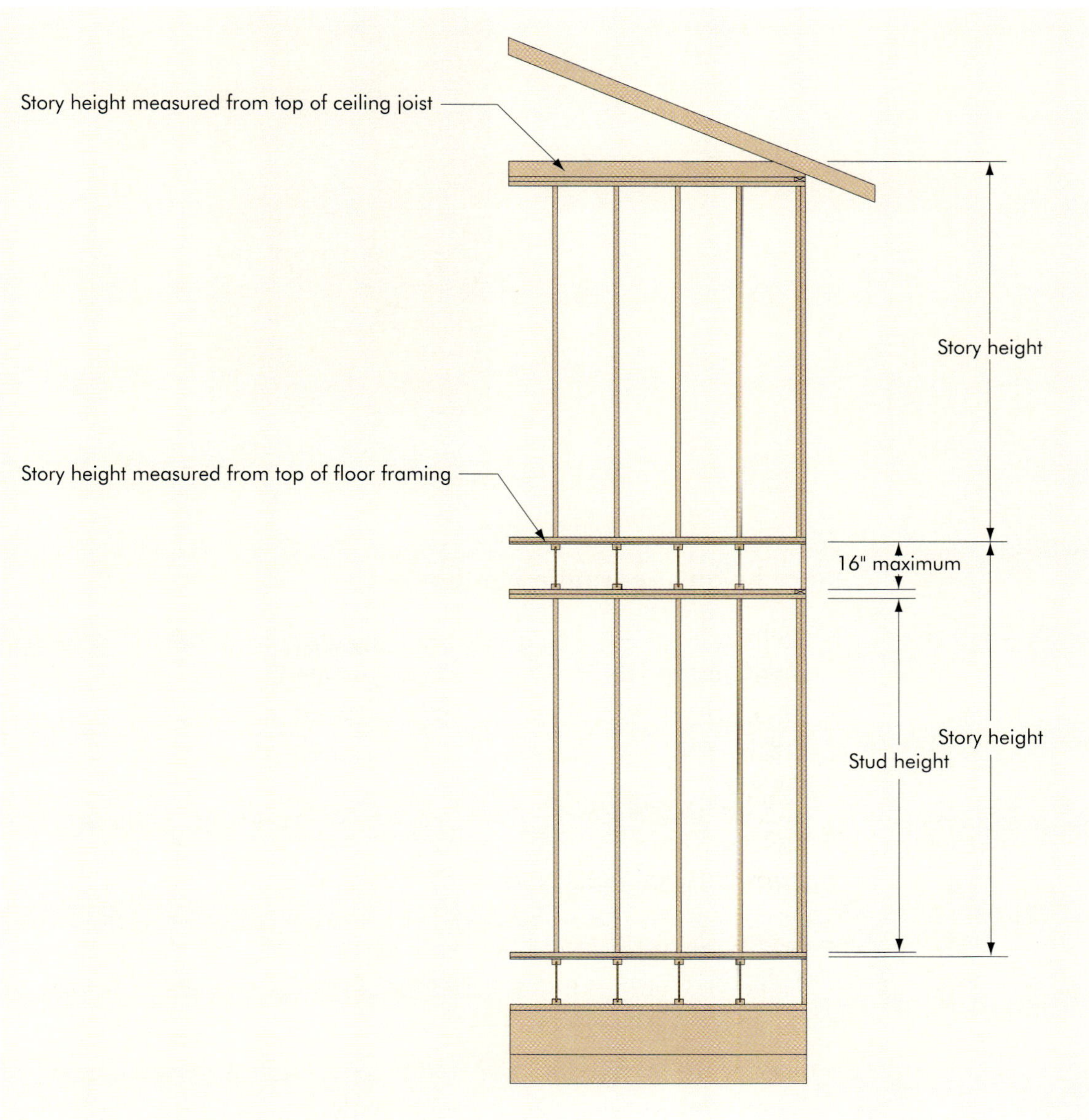

FIGURE 4.3

STORY HEIGHT MEASUREMENT FOR MASONRY FRAMING

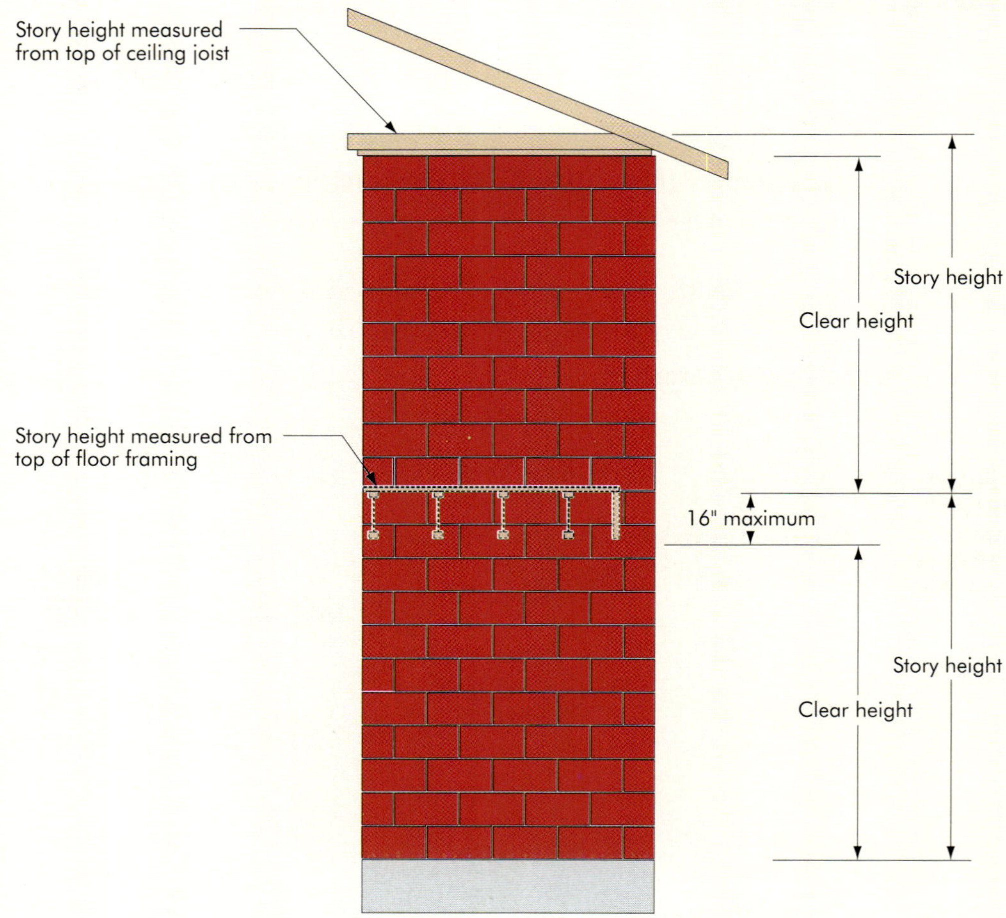

TABLE 4.1

Story heights[a] permitted in the IRC

Wall Construction	Maximum Height (ft)	Maximum Floor Framing Height (in.)
Wood	10[b]	16
Steel framing	10	16
Masonry	12[c]	16
Insulated concrete form	See Section R611	16

a. Laterally unsupported bearing wall heights.

b. Wall height shall be permitted to be increased to 12 ft provided the bracing amounts are increased by multiplying by a factor of 1.20 in accordance with Section R301.3, item 1, Exception.

c. An additional 8 ft shall be permitted at gable end walls.

In addition to the overall story height, this code section also limits the floor framing depth to a maximum of 16 inches and the stud length to 10 feet. For wood-framed residences, the stud length can be increased to 12 feet provided the amount of bracing required by IRC Table

R602.10.1 is increased by a factor of 1.2. See Table 4.2 for bracing adjustments for wood-framed walls. While not explicitly provided for in this section, linear interpolation of the bracing factor for stud lengths between 10 and 12 feet is reasonable and justifiable from an engineering perspective.

Question: Can a floor depth greater than 16 inches be used as long as I do not exceed my maximum story height of 136 inches (120 + 16)?

Answer: While not covered in the current version of the IRC, this question has come up a number of times. From an engineering perspective, the load contributed by a wall 136 inches high is essentially the same regardless of how much of the 136 inches in overall height is stud length and how much is floor-framing depth. Therefore, for this example, using a 24-inch-deep floor joist should be permissible as long as the stud length is 112 inches or less (136 – 24).

TABLE 4.2

Bracing adjustments required for wood-framed walls greater than 10 feet

Wall Height (ft)	Bracing Amount Factor
10 or less	No factor
>10 to 12	1.2

4) Design criteria – what are the local design requirements?

Every jurisdiction that adopts the IRC is required to fill out Table R301.2(1), shown below in Table 4.3. The IRC user will need to get this information from the jurisdiction where the building will be constructed. As the IRC is not a one-size-fits-all document, there are a number of design criteria the user must know for the project that Table R301.2(1) summarizes. For example, in order to determine the project's bracing requirements, the designer must know the SDC and wind speed for the location.

TABLE 4.3

Local jurisdictions must specify climatic and geographic criteria.

TABLE R301.2(1)
CLIMATIC AND GEOGRAPHIC DESIGN CRITERIA

Ground Snow Load	Wind Speed (mph)	Seismic Design Category	Subject To Damage From			Winter Design Temp	Ice Barrier Underlayment Required	Flood Hazards	Air Freezing Index	Mean Annual Temp
			Weathering	Frost Line Depth	Termite					

5) Wind loads

Section R301.2.1 of the code defines the external wind pressures that components of a structure must be able to resist.

Wind acts on the exterior components of a building. The resulting load on these components is then transferred to the main structural elements of the building. These components are listed in Section R301.2.1 and include exterior wall coverings, curtain walls, roof coverings, exterior windows, skylights, doors and veneer. A person familiar with the IBC would immediately recognize these items as "components and cladding." In the IBC the main structural elements that provide lateral support for the building are called "main wind-force-resisting systems."

In the IRC, main wind-force-resisting systems are treated prescriptively via specified minimum roof and floor sheathing and attachment schedules, and through the required wall bracing. As such, by selecting the appropriate levels of bracing specified in R602.10.1, and following the limitations specified in Chapter 3 of the IRC, the user will meet the necessary requirements for main wind-force-resisting systems without design or further consideration.

IRC Table R301.2(2) gives the "component and cladding" pressure requirements that must be met. In Table R301.2(2) pressure values are given for basic wind speeds in 5 mph increments based on the wind speed, roof slope, and effective wind area for the various "zones" of the structure (as defined in Figure R301.1(7)). These zones are the locations to which the pressure values are applied. Zones 1, 2 and 3 apply to roofs and Zones 4 and 5 apply to walls. The footnote to Figure R301.1(7) specifies a 4-foot dimension for the width of the end zones at the corners and edges of the structure. Buildings having roof slopes exceeding the maximum slopes must be engineered using the *International Building Code* per IRC R301.1.

The "effective wind area" term used in the table is not defined in the IRC. A definition, however, is provided in ASCE 7 – Minimum Design Loads for Buildings and Other Structures.

The effective wind area for component and cladding elements is the span length multiplied by an "effective width." This "effective width" need not be less than one-third of the span length. For cladding fasteners, the effective wind area shall not be greater than the area that is tributary to an individual fastener.

EXAMPLE:

What is the effective wind area for a piece of 4- by 8-foot wall sheathing applied vertically on an 8-foot-tall wall? As the wall studs span 8 feet, the effective width is 8' ÷ 3 = 2.67'. The effective area need not be less than 8' x 2.67' = 21.3 square feet. Thus an effective wind area of 20 feet would be reasonable for use in IRC Table R301.2(2).

It can be seen from Table R301.2(2) that wall pressures for components and cladding range from about +13 to –27 psf for the 85 to 105 mph basic wind speed in Exposure B, Zone 5. The pressure from this table is then modified by a "height and exposure" factor from Table R301.2(3) to determine the design pressure. The exposure categories, ranging from A to D, are defined in Section R301.2.1.4, and discussed later in this chapter. The exterior wall cladding components (wall coverings, curtain walls, roof coverings, exterior windows, skylights, doors) **must** have sufficient capacity to withstand these wind pressures.

Component pressure capacities can be obtained from wall and roof covering manufacturers and sheathing manufacturers. It cannot be assumed that a product meets the code pressure requirements just because the product is commonly used. Given the performance of a number of siding and sheathing products during the 2005 hurricane season in the Southeast, it appears that many siding and sheathing products have insufficient capacity to resist the wind pressure requirements in some parts of the U.S.[a,b,c] Some exterior components may require additional fastening or closer support spacings to meet the requirements of this section.

a. NIST Technical Note 1476. *Performance of Physical Structures in Hurricane Katrina and Hurricane Rita: A Reconnaissance Report.* National Institute of Standards and Technology, Gaithersburg, MD 20899. p. 154, 158, 170. http://www.bfrl.nist.gov/investigations/pubs/NIST_TN_1476.pdf

b. APA Form SPE-1125. *Hurricane Katrina – Structural Performance of Wood Frame Buildings in the Aftermath.* APA – The Engineered Wood Association, Tacoma, WA 98466. p. 5–8.

c. van de Lindt, J.W., A Graettinger, R. Gupta, T. Skaggs, S. Pryor, and K. Fridley. (2007). Performance of "Woodframe Structures During Hurricane Katrina". *ASCE Journal of Performance of Constructed Facilities,* March/April 2007, Volume 21, Number 2, p. 108.

The example below shows the steps to determine the component and cladding wall pressures.

EXAMPLE:

What is the wind suction pressure for the wall sheathing panel in the previous example? The panel is in a single story home and located in a 90 mph wind zone, Exposure B.

The worst case for a wall panel is its placement at the corner of the building. This would put it in Zone 5 (see Figure R301.1(7)). From Table R301.2(2) – Wall, Zone 5, 20 square feet, 90 miles per hour – values of +13.9 psf, and –18.2 psf are shown. The positive (+) value is for wind load acting from the outside toward the inside of the house (pushing the panel toward the wall studs). The negative (–) value is the wind load pulling the panel away from the studs, creating suction. In answer to the question, the wind suction pressure is 18.2 psf.

Notice that there are two wind pressure values given in the example above. This is because as the orientation of the wind changes with respect to the structure, the wind can cause either suction or positive pressures on the structure. For example, a wind blowing on the north side of a structure causes a positive pressure on the north side and simultaneously causes negative pressure or suction on the south side. An element must be built to resist both pressures because the wind may blow from any direction.

Construction in high wind areas (R301.2.1.1)

If the wind speed limit for the IRC is exceeded (wind speed 110 mph or more, or 100 mph or more in hurricane-prone regions) design documents listed in this section must be used, as shown in Table 4.4.

TABLE 4.4

Applicable design documents

R301.2.1.1 Design Criteria

Basic Wind Speed	Applicable Document
Less than 110 mph	IRC
100 mph or more in Hurricane-Prone Regions[a]	See R301.2.1.1 for design reference document
110 mph or more	

a. IRC Section R202 – Definitions: Hurricane-Prone Regions. Areas vulnerable to hurricanes, defined as the U.S. Atlantic Ocean and Gulf of Mexico coasts where the basic wind speed is greater than 90 mph (145 km/h), and Hawaii, Puerto Rico, Guam, Virgin Islands, and American Samoa.

Exposure categories (R301.2.1.4)

Exposure categories take into account the shielding effects on wind load from other buildings, natural topography and vegetation. Adjacent buildings can deflect wind, thus reducing the pressure on the building. Tall vegetation, such as trees and terrain with numerous, closely spaced obstructions (the size of single-family houses or larger), can also reduce the effect of wind. Topographical features such as flat, open areas are associated with higher exposure categories because there are no obstructions to reduce the effects of the wind. The code defines four wind exposure categories ranging from A to D. The category that results in lowest wind pressures is exposure category "A" and the highest is "D."

Exposure Category B shall be assumed (R301.2.1.4) unless otherwise directed by Table R301.2(1) in the IRC adopted in your jurisdiction.

See Figures 4.4 through 4.10 along with the exposure category descriptions in the IRC to determine the appropriate wind exposure category for a building's location.

FIGURE 4.4
EXPOSURE CATEGORY A

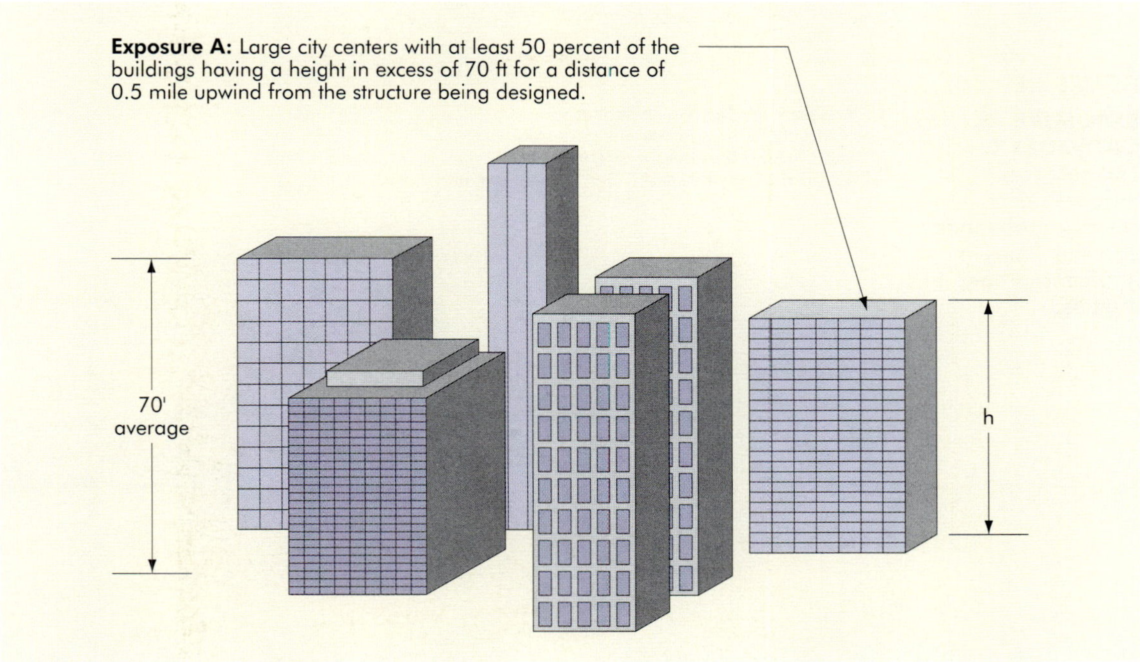

FIGURE 4.5

EXPOSURE CATEGORY B – *Shall be* ***assumed*** unless the site meets the definition of another type exposure.

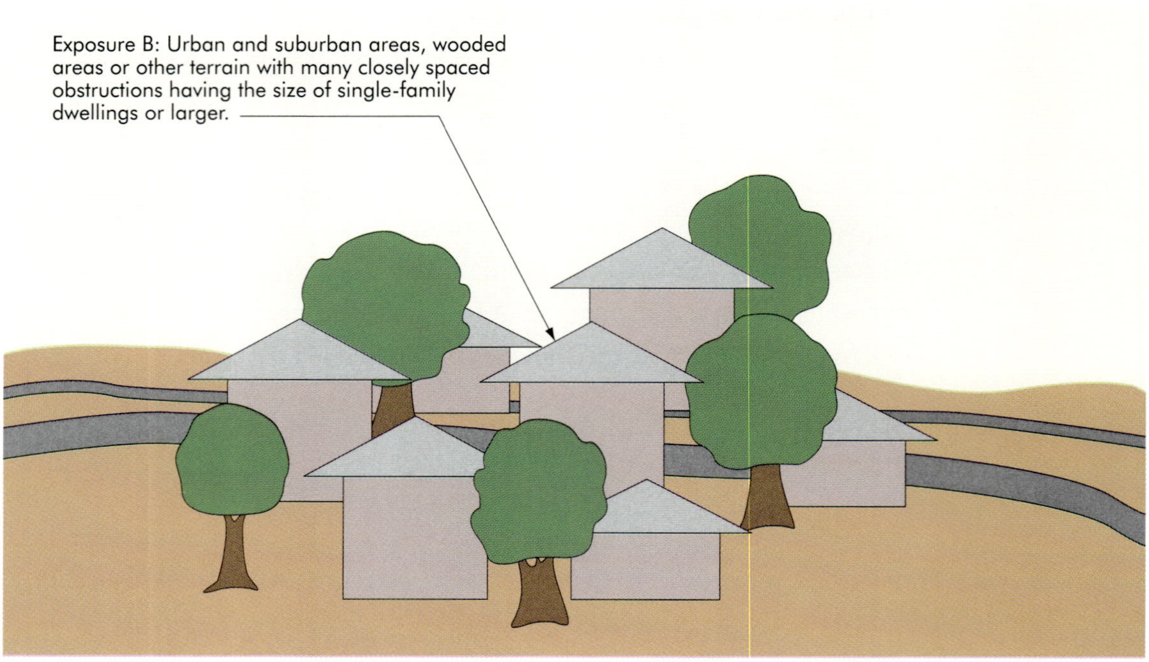

FIGURE 4.6

EXPOSURE CATEGORY C – This category includes open country, grasslands and shorelines in hurricane-prone regions.

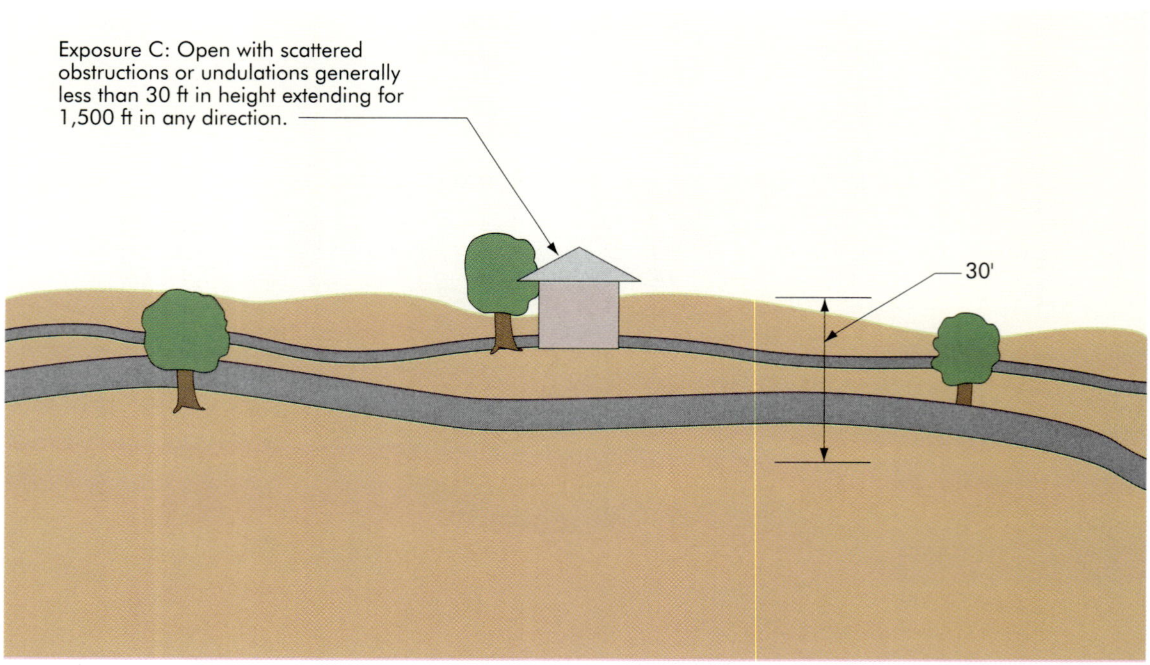

FIGURE 4.7

EXPOSURE CATEGORY C (Continued) – This category includes open country, grasslands and shorelines in hurricane-prone regions.

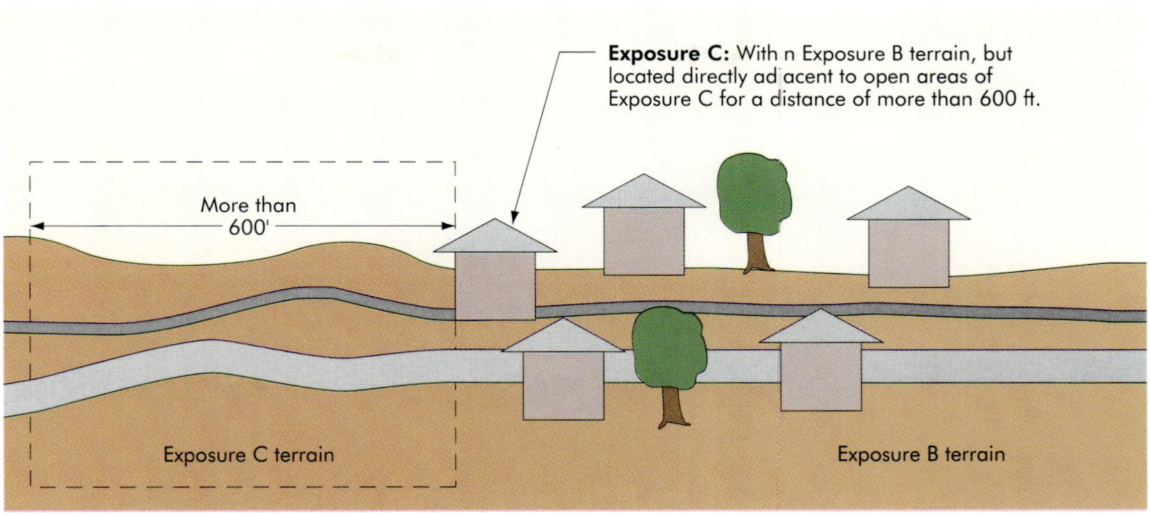

FIGURE 4.8

EXPOSURE CATEGORY D

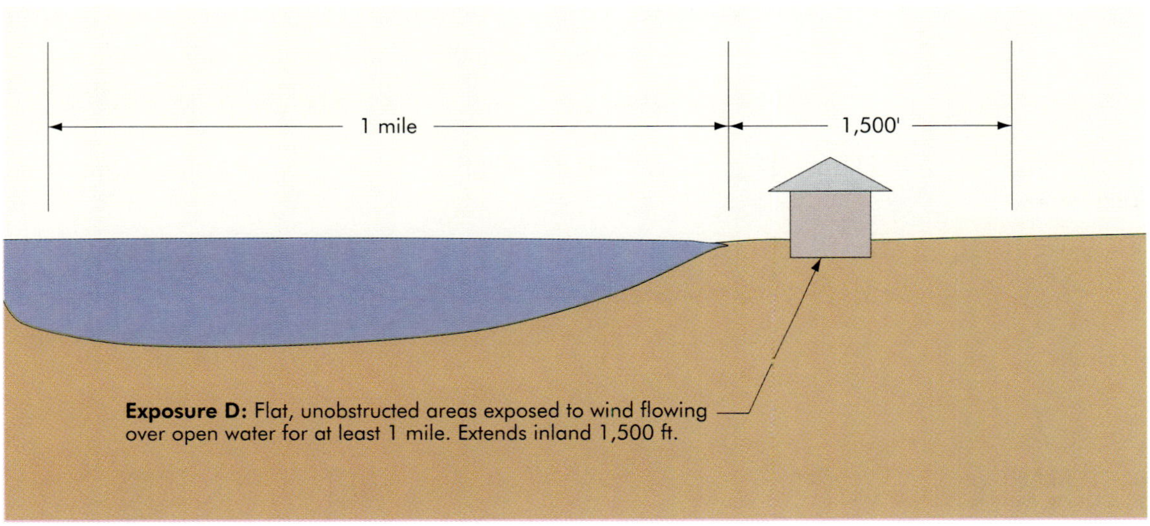

6) Seismic loads – Seismic Design Categories A–C

Seismic exemptions –

Section R301.2.2 provides the limitations applicable to structures designed for the seismic requirements of the IRC. One of the most important and most often missed provisions is that the application of the seismic requirements is different for one- and two-family dwellings than for buildings with three or more attached dwelling units (townhouses). IRC Section R202 gives the definitions of dwelling unit, dwelling and townhouse.

> ***DWELLING UNIT.*** *A single unit providing complete independent living facilities for one or more persons...*
>
> ***DWELLING.*** *Any building that contains one or two dwelling units...*
>
> ***TOWNHOUSE.*** *A single-family dwelling unit constructed in a group of three or more attached units in which each unit extends from the foundation to roof and with open space on at least two sides.*

In Figure 4.9, plans "A" and "B" meet the requirements of a townhouse because at least two sides of each unit have open space – meaning exterior walls. The plan shown in Figure 4.10 does not meet the townhouse definition because the middle units do not have two sides that are open (having exterior walls). The middle units only have one side that is open. All units of a townhouse must extend from the foundation to the roof. This means that dwelling units cannot be stacked one above another.

FIGURE 4.9

BOTH PLANS A AND B MEET TOWNHOUSE REQUIREMENTS BECAUSE EACH UNIT HAS OPEN SPACE ON TWO SIDES

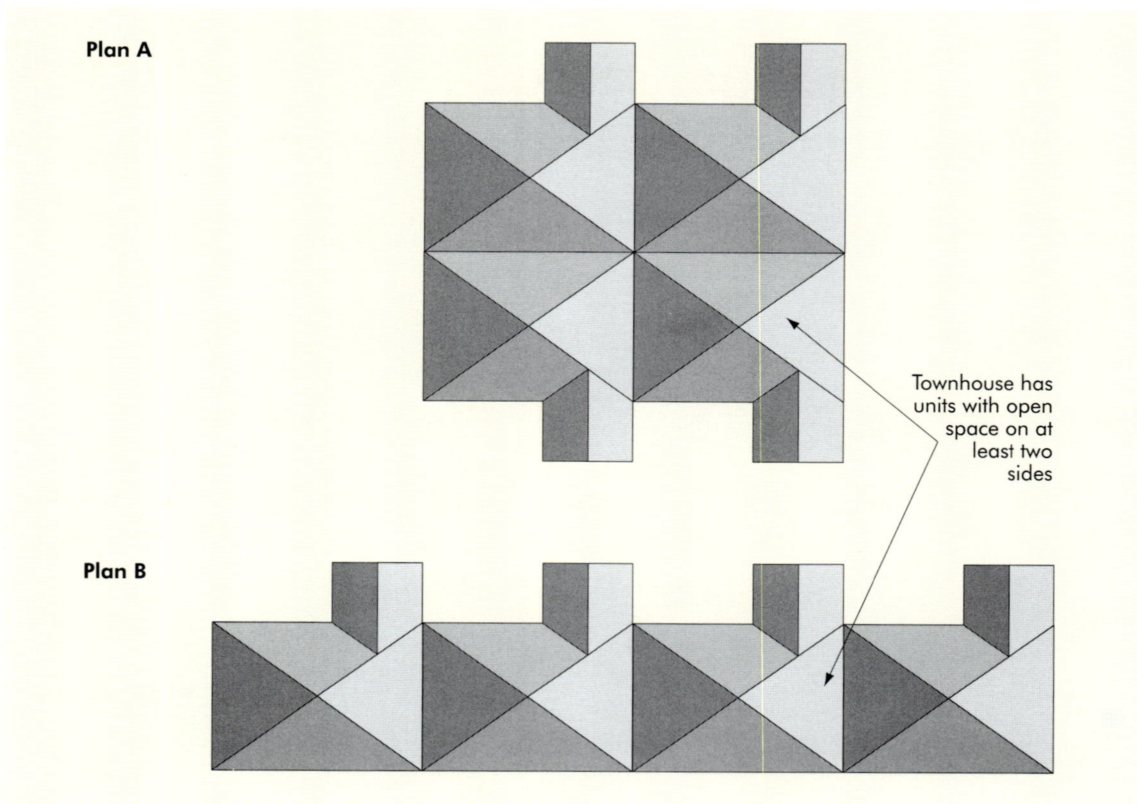

FIGURE 4.10

PLAN VIEW SHOWS A CONFIGURATION NOT COVERED BY THE IRC

Structure does not meet requirements of a dwelling (more than 2 units) or a townhouse (middle units not open on two sides).

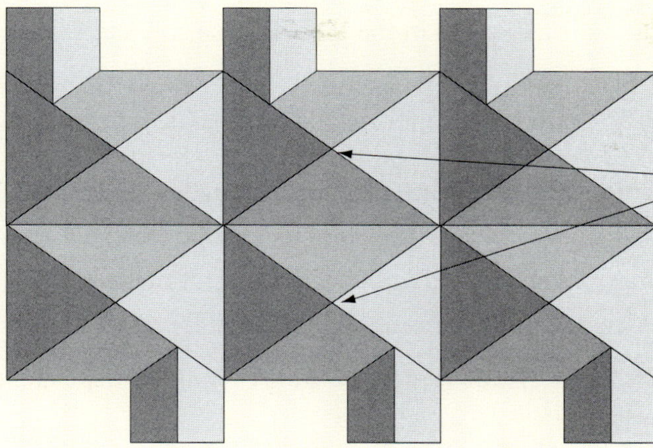

Table 4.5, developed from Section R301.2.2, illustrates the importance of the distinction between dwellings and townhouses. Note that *dwellings* located in Seismic Design Categories A, B, and C are **exempt** from the seismic requirements of the IRC. This means that these *dwellings* are not required to comply with the seismic provisions of 301.2.2 and all of its subsections as well as those in Chapters 4, 5, 6 and 7. This exception significantly reduces the construction requirements for one- and two-family dwellings built in Seismic Design Categories A, B, and C. However, *dwellings* in Seismic Design Categories D_0, D_1, and D_2 and *townhouses* in any SDC must comply with the seismic requirements of R301.2.2.

TABLE 4.5

Scope of seismic provisions

Seismic Design Categories	Seismic Requirements	
	Dwellings	Townhouses
A – C	Exempt	No Exemption
$D_0 - D_2$	No Exemption	No Exemption

IRC seismic basics

The following is a discussion of the basic concepts, along with the corresponding IRC section numbers, needed to apply the IRC seismic provisions.

Determination of seismic design category (R301.2.2.1)

The SDC is determined from the information located in Table R301.2(1) of your locally adopted IRC. If not available, the map shown in IRC Figure 301.2(2) may be used, or contact your local building official. Some jurisdictions may amend the local code to require a different seismic design category than that specified by the IRC seismic map.

Alternate determination of seismic design category (R301.2.2.1.1)

The SDC used in the IRC, as well as in the IBC, assume soil Site Class D when the soil properties are not known in sufficient detail to determine the site classification. See 2006 IBC Section 1613.5.5 or ASCE7-05 Section 11.4.2 for details on Site Class determination.

Alternative determination of Seismic Design Category E (R301.2.2.1.2)

There are some cases where a building located in Seismic Design Category E, according to Figure R301.2(2), may be reclassified as Seismic Design Category D_2. This would put them within the scope of the IRC, thus eliminating the need for a full design. These cases are described in R301.2.2.1.2.

Seismic limits – Seismic Design Categories A–D_2

The following seismic limitations apply to residential structures located in all SDCs. In subsequent sections, **additional** limitations for Seismic Design Categories C and D are discussed.

Seismic limitations and weight of materials (R301.2.2.2)

The lateral seismic loads on a building are related to building weight (mass). The greater the building weight, the greater the seismic load (remember, force = mass x acceleration). During ground motion, the structural parts of a building must resist the lateral forces to prevent collapse. Therefore, the IRC sets limits on the weights of floors, walls, ceiling and roofs. Table 4.6 lists the maximum building weights defined by this section for wood frame construction. Some building weights can be increased if the amount of wall bracing is increased. Increasing the amount of wall bracing makes the building stronger, stiffer and more capable of resisting the seismic force caused by greater weight.

TABLE 4.6

Bracing adjustment factors based on weights of construction materials (per Section R301.2.2.2.1)

Dead Loads and Wall Adjustment Factors			
Component	Maximum Dead Load Permitted for Standard Wall Bracing (psf)	Maximum Dead Load Permitted with Wall Bracing Adjustment (psf)	Bracing Panel Amount Adjustment Factor[a]
Floor	10	N/A	N/A
Wall Dead Load, Interior	10	8	0.85
Wall Dead Load, Exterior	15	8	0.85
Roof/Ceiling Dead Load For Wall Supporting the Roof Only	15[b]	25[b]	1.1
Roof/Ceiling Dead Load For Wall Supporting the Roof Plus One Story	15[b]	25[b]	1.2

a. This factor is to be applied to the required bracing amount (length or percent) in the wall line. From IRC Table R301.2.2.2.1 and Table R602.10.1, footnote d.

b. The allowable load given for roofs is the weight of the roof based on the horizontal projection of the roof. This is the total weight of the roof divided by the area of the roof in plan view.

Example: Consider a braced wall that has only a roof above it. The roof is heavy tile with a total dead load of 20 psf measured on the horizontal projection. Since 20 psf exceeds the 15 psf limit, but is less than 25 psf, the amount of bracing required per R602.10.1 must be increased by a factor of 1.1.

EXAMPLE:

A home designer draws up a set of plans for a two-story townhouse. After working through the bracing section of the IRC, the designer determines that the structure requires each exterior wall on the first story to have 25% bracing and the second story to have 16%. The designer assumes that, like most townhouses in the area, it will have a typical lightweight wood roof (roof/ceiling dead load of 15 psf or less); however, the owner later asks the designer to consider a slate roof. This roofing material adds 6 psf to the weight to the horizontal projection of the roof. How does this impact the percentage of bracing needed for the first and second stories?

With a slate roof, the weight of the roof/ceiling becomes 21 (15+6) psf. From Table 4.5, 21 psf is less than the maximum dead load permitted – 25 psf. This means that a bracing length adjustment factor of 1.2 is required for the first story and 1.1 for the second story.

First story 25% x 1.2 = 30% bracing required

Second story 16% x 1.1 = 17.6% bracing required

If the designer had already figured out the lengths of bracing required for each wall on each story, the adjustment factors could have been applied to the wall lengths. It is only applied once, either to the percentages or the lengths.

FIGURE 4.11

MAXIMUM DEAD LOAD WEIGHTS FOR WOOD ASSEMBLIES

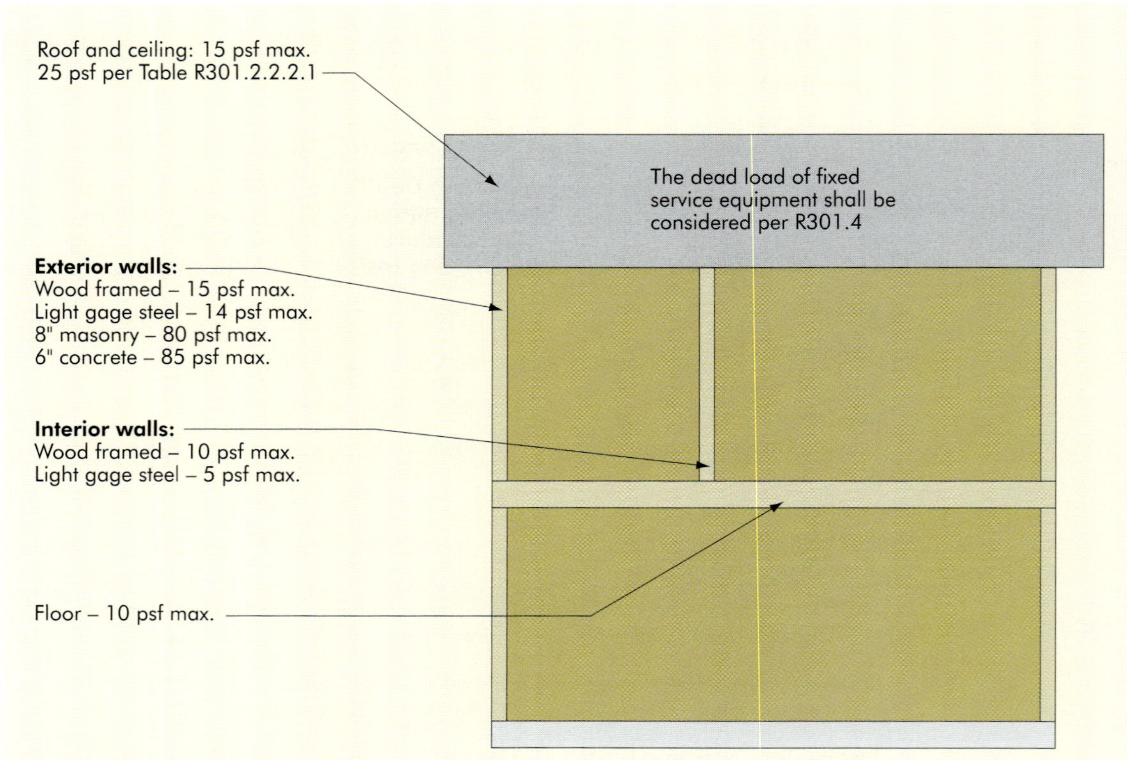

Note: The allowable load given for roofs is the weight of the roof based on the horizontal projection of the roof. This is the total weight of the roof divided by the area of the roof in plan view.

Building irregularities (R301.2.2.2.2)

This section of the code defines building shapes that are *irregular.* It is important to note that this code section does not apply to all structures covered by the IRC. See Table 4.5 for the applicability of the seismic provisions for dwellings and townhouses. The IRC seismic provisions assume that a building has a relatively uniform shape. If the building deviates too much from the assumed uniform shape, the building or portion of the building is considered irregular.

The problem with *irregular* buildings is that the irregular shape of the structure causes the lateral loads to get shifted around in the building in unusual and difficult-to-predict load paths. Such structures are difficult to design and almost impossible to economically specify using the prescriptive provisions of the IRC. In order to permit some degree of building irregularity that the prescriptive provisions can safely accommodate, this section was developed. Those structures that are outside of these limitations are *irregular* shaped buildings, or portions of buildings, and must be designed using the IBC to ensure that the building has an adequate load path to resist seismic loads. Buildings having a *regular* shape can be constructed by following the prescriptive provisions of the IRC.

Many contemporary home designs may be considered irregular by one of the seven conditions of this section. These provisions, however, include some exceptions that may permit an otherwise *irregular* home to be constructed by the IRC.

Irregularity #1. Braced wall panels are not in a vertical plane

The first irregularity provision requires that braced wall panels be vertically aligned in one plane.

FIGURE 4.12

IRREGULARITY #1
Braced wall panels with vertical irregularities.

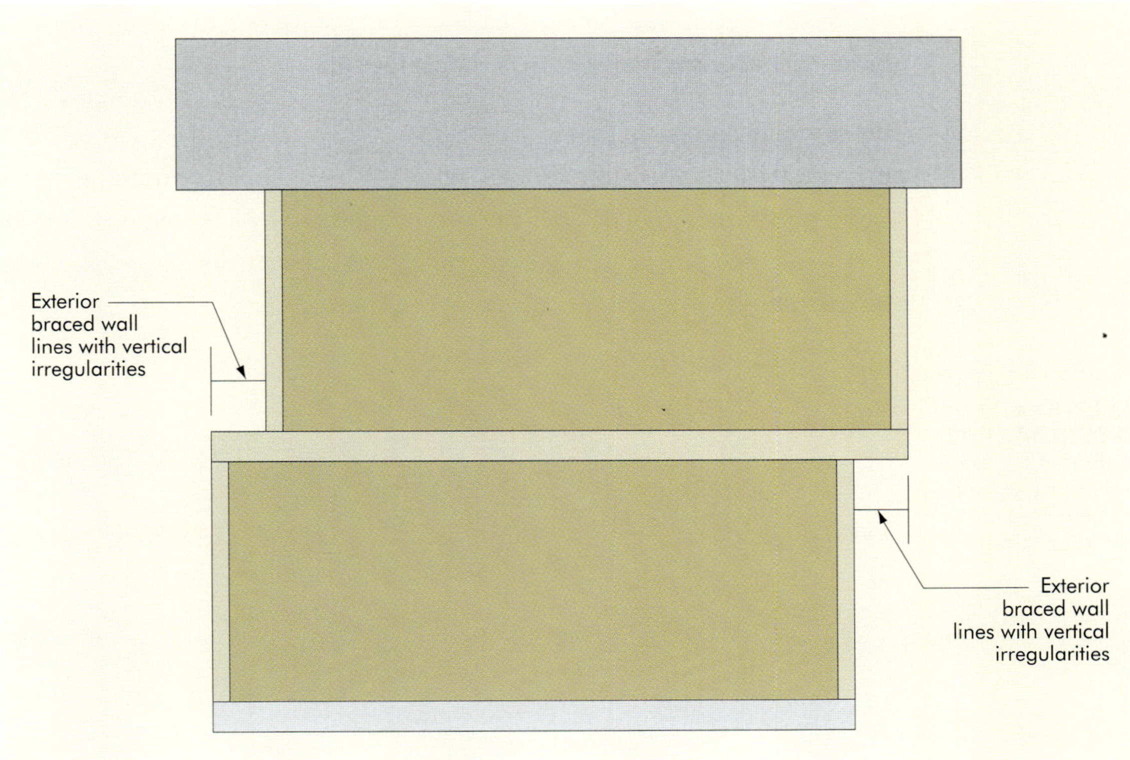

Following this irregularity provision, an exception is listed for buildings constructed using wood light-frame construction as shown in Figure 4.12. This exception is not applicable to other types of construction, such as masonry and concrete. According to the exception, braced wall panels do not have to be in one plane vertically if:

Setbacks or cantilevers do not exceed 4 times the nominal depth of the wood floor joists and *ALL* of the following conditions are met:

1. Floor joists are nominal 2x10 or larger and spaced not more than 16" o.c.
2. The ratio of the backspan to cantilever is at least 2:1.
3. Floor joists at the ends of braced wall panels are doubled.
4. A continuous rim joist is connected to ends of all cantilevered joists.
5. Gravity loads at the end of the cantilever are limited to uniform wall and roof loads and the headers in the wall must have a span of 8 feet or less.

Irregularity #2. Braced wall line not under floor or roof

According to the second provision, a building is considered irregular when a braced wall line is not beneath (or supporting) a portion of a roof or floor. If this section of floor is not in turn supporting any braced wall panels from above, it is permitted to extend out not more than 6 feet. The exception to the irregularity is shown in Figure 4.13.

This second irregularity provision is different than the first irregularity. The first irregularity provision relates to braced wall lines being supported by floor framing. In the exception to the second provision, a roof or floor can be unsupported by a braced wall line for a maximum of 6 feet. This code provision relates to light-frame construction as well as concrete and masonry construction.

FIGURE 4.13
EXCEPTION TO IRREGULARITY #2
Portions of floor or roof not supported are permitted to extend up to 6 ft.

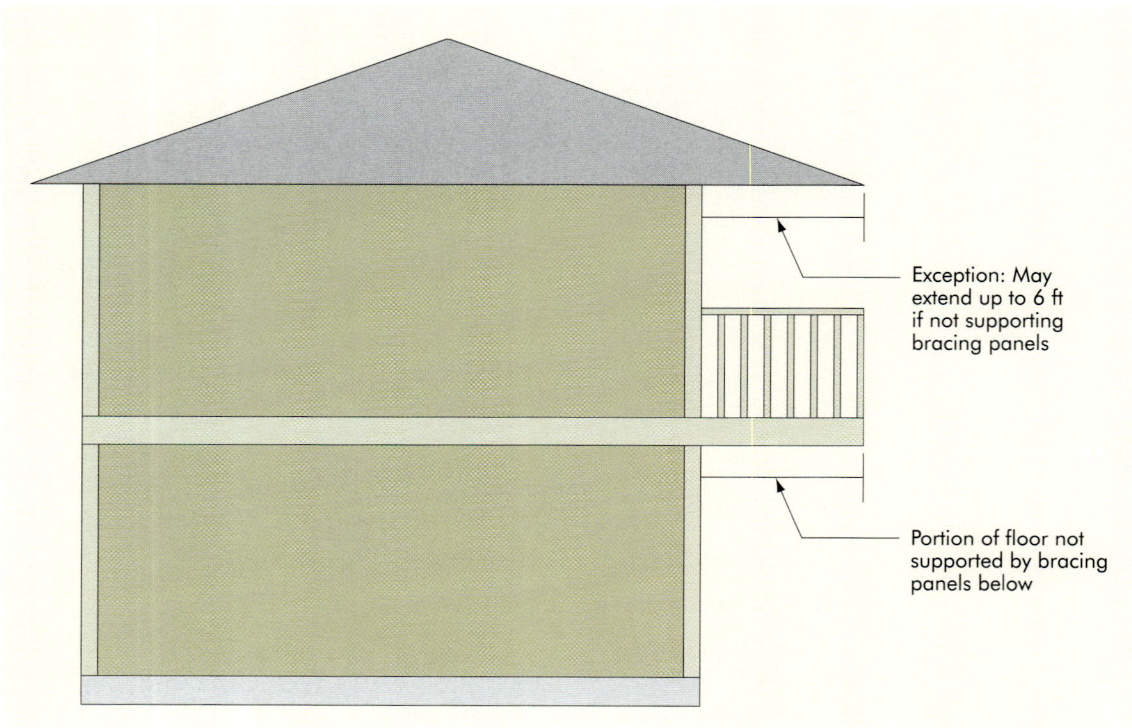

Irregularity #3. Braced wall panel occurs over opening below

The third irregularity provision in this section places limitations on the location of the end of braced wall panels relative to openings in the braced wall line below. For steel light-frame construction, a braced wall panel is only permitted to extend 1 foot over an opening in a lower wall line as shown in Figure 4.14. For wood light-frame construction, a braced wall panel is permitted to extend more than 1 foot over the opening below provided the entire braced wall panel is not located over the opening. When a wood light-framed braced wall panel extends more than 1 foot over an opening below, the additional header requirements listed in Table 4.7 must be met.

FIGURE 4.14
IRREGULARITY #3
Braced wall panel over openings.

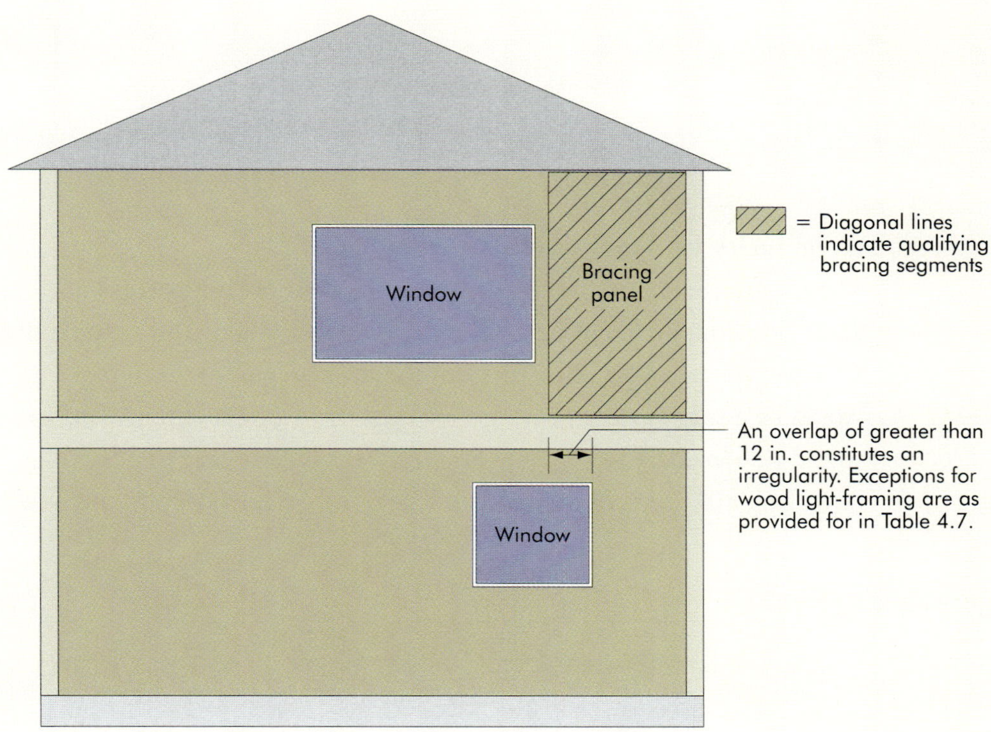

For wood light-frame wall construction, a number of exceptions apply to irregularity #3. Provided that the building width, loading conditions, and framing member species meet the requirements of Table R502.5(1), and the entire length of the braced wall panel does not occur over the opening, the header sizes shown in Table 4.7 may be used to exempt irregularity #3.

TABLE 4.7

Header requirements to exempt irregularity #3

Maximum Window Opening Width Below Bracing	Minimum Header Requirements
4 ft	1 – 2x12 2 – 2x10
6 ft	2 – 2x12 3 – 2x10
8 ft	3 – 2x12 4 – 2x10

Irregularity #4. Floor or roof opening limit

The fourth irregularity limits the size of the floor opening between stories or an opening in a roof. This limitation applies to all types of construction. The purpose of this restriction is to limit the size of a hole in the floor or roof diaphragm. An example of an opening between floors would be a stairway or room with a height that extends through the story above. A skylight would be considered a roof opening. The size of floor and roof openings is limited to the lesser of 12 feet or 50 percent of the least floor or roof dimension as shown in Figure 4.15.

FIGURE 4.15
IRREGULARITY #4
Excessive hole in roof or floor sheathing.

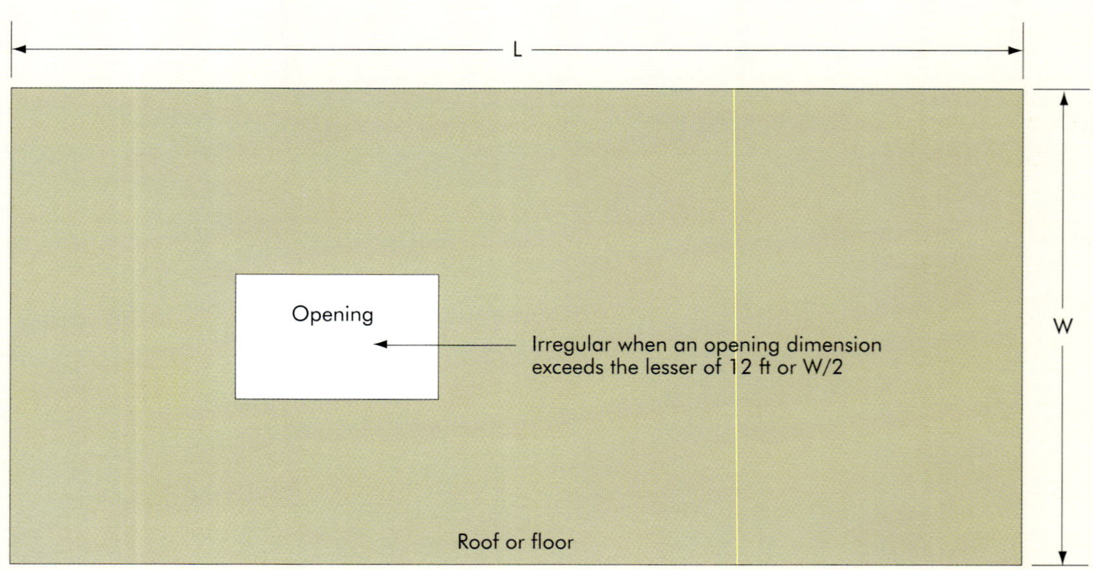

Irregularity #5. Vertical offsets in floor level

For the fifth irregularity provision, a building is considered *irregular* when a floor has a vertical offset as shown in Figure 4.16. For wood light-framed construction, the building is considered *regular* if the floor framing is lapped or tied together per IRC section R502.6.1. For all construction types (wood, steel, etc.), vertical offsets in a floor level shall not be considered irregular if the perimeter framing of the floor is supported directly on a continuous foundation.

Code section R502.6.1 states that joist framing must be lapped a minimum of 3 inches and face-nailed together with three 10d nails. As an alternate, a wood or metal splice with equal strength shall be permitted. These measures address the problems associated with these irregularities.

FIGURE 4.16
IRREGULARITY #5
Offset in floor framing.

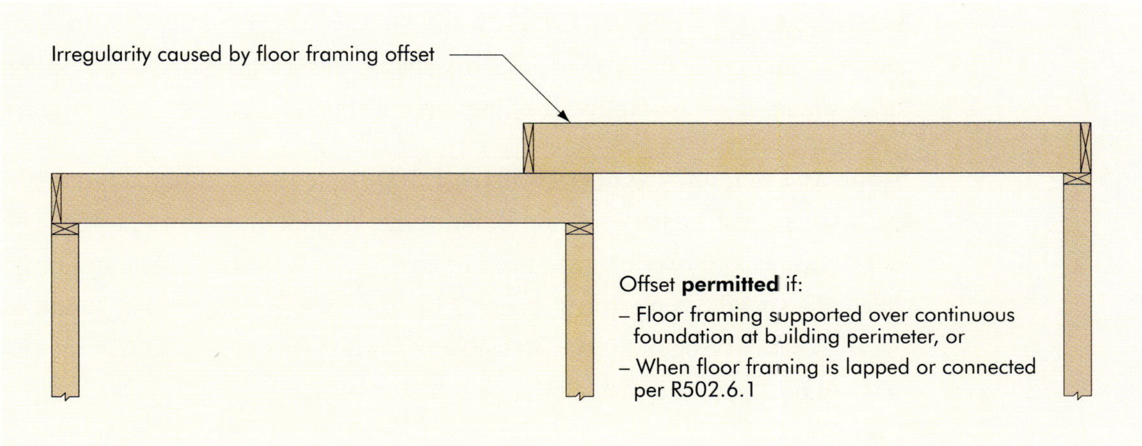

Irregularity #6. Braced wall lines must occur in perpendicular directions

Irregularity provision #6 simply states that if braced wall lines are not perpendicular to each other, the building must be considered *irregular*. Figure 4.17 illustrates such an irregularity.

FIGURE 4.17
IRREGULARITY #6
Braced wall lines not at right angles to each other.

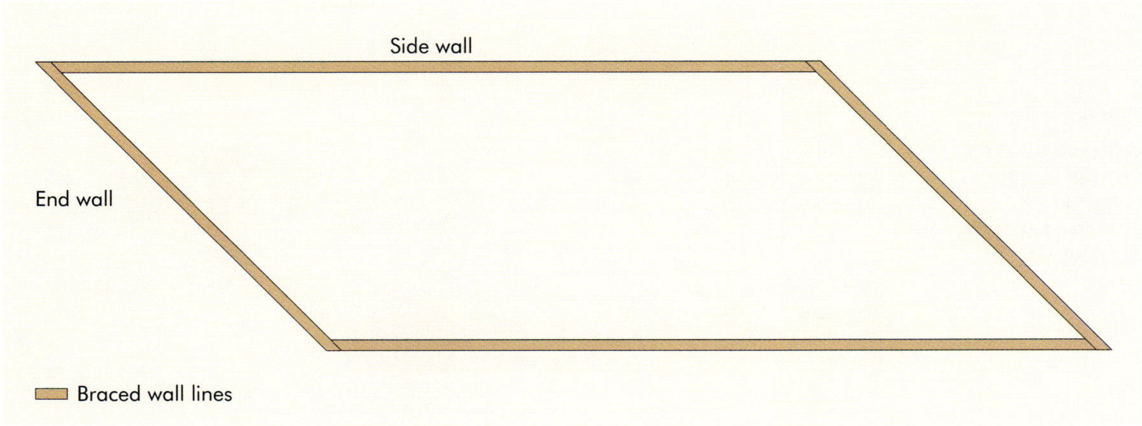

Irregularity #7. Above-grade masonry or concrete construction
The seventh irregularity provision states that wood- and steel-framed buildings braced in accordance with IRC Chapter 6 are irregular when they include above-grade masonry or concrete construction. Fireplaces, chimneys and masonry veneer are permitted provided they are constructed in accordance with other provisions of the IRC. Note that when this irregularity exists, the entire story shall be designed and not just the portion causing the irregularity.

Additional seismic limits – Seismic Design Category C and above

In addition to the limitations discussed above, there are additional limitations placed on some residential structures located in Seismic Design Category C and above:

Seismic Design Category C (R301.2.2.3). This section of the code deals with veneer, masonry and concrete in structures in Seismic Design Category C. As discussed previously, since single- and two-family dwellings are exempt, this applies to townhouses only.

Stone and masonry veneer (R301.2.2.3.1). Note that additional bracing is not required in Seismic Design Categories A and B. Buildings located in Seismic Design Category C having stone and masonry veneer may need to have an increase in the amount of wall bracing provided. The height of masonry veneer is limited to 30 feet above a noncombustible foundation. An additional 8 feet above the 30 feet is permitted for gable end walls. See Table 4.8 for a summary of the stone and masonry veneer requirements.

When stone or masonry veneer is used on multiple story buildings, all stories except the top story are required to have the amount – length or percent – of bracing per IRC Table R602.10.1 multiplied by a factor of 1.5 on both interior and exterior walls. For all SDCs, light wood-framed buildings with stone or masonry veneer are limited to three stories and light steel-framed buildings are limited to two stories. See Figure 4.18 for an illustration. See IRC Section R703.7 and Table R703.7(1) for provisions for the height requirements for stone and brick veneer over wood and steel light-framed walls.

FIGURE 4.18
SUMMARY OF STONE AND MASONRY REQUIREMENTS OF R703 FOR SDC C

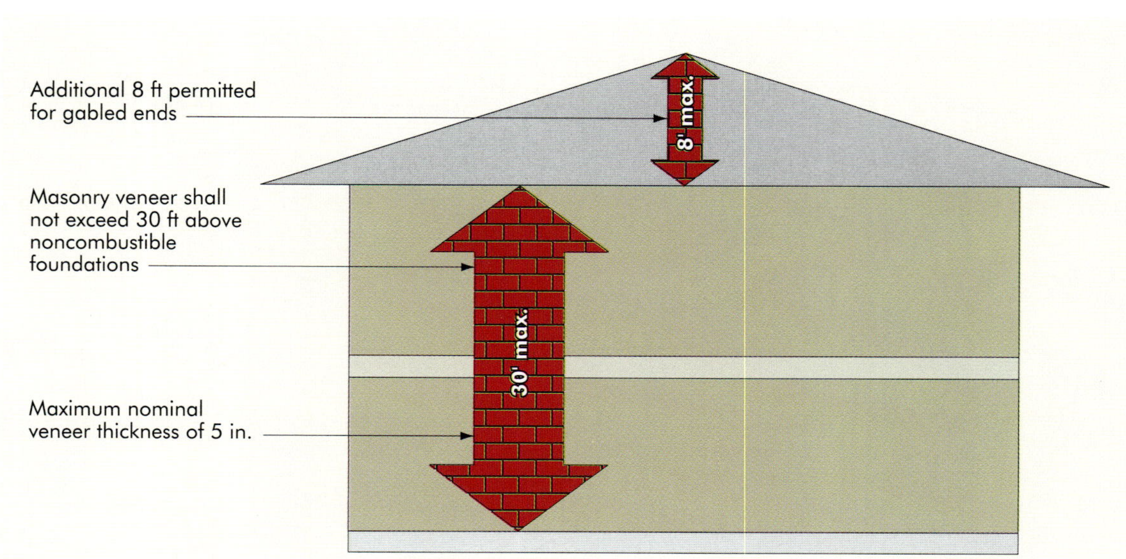

TABLE 4.8

Stone or masonry veneer minimum bracing requirements for detached one- and two-family dwellings

Seismic Design Category (SDC)	Number of Framed Stories	Framed Story	Factor to be Multiplied by Length Required in Table R602.10.1
A or B	1, 2 or 3	All	1.0
C	1, 2 or 3	Top	1.0
		Middle	1.5
		Bottom	1.5

Additional seismic limits – Seismic Design Categories D_0, D_1 and D_2

In addition to the limitations just discussed, there are additional limitations placed on residential structures in Seismic Design Categories D_0, D_1 and D_2:

Seismic Design Categories D_0, D_1 and D_2 (R301.2.2.4). Buildings constructed in Seismic Design Categories D_0, D_1 and D_2 are required to meet all of the provisions of buildings constructed in Seismic Design Category C plus the added requirements of this section, which has five subsections.

Height limitations (R301.2.2.4.1). This section defines the maximum number of stories permitted for light wood- and steel-framed buildings in Seismic Design Categories D_0, D_1 and D_2. Light wood-framed buildings are permitted to be a maximum of three stories.

TABLE 4.9

Height limitations (Seismic Design Categories D_0, D_1 and D_2)

Framing Material	Maximum Number of Stories
Wood	3[a]

a. Table R602.10.1 limits wood-framed structures in SDC D_2 to 2 stories and a cripple wall.

Stone and masonry veneer (R301.2.2.4.2). The stone and masonry veneer limits for Seismic Design Categories D_0, D_1, and D_2 are more stringent than they are for Seismic Design Category C:

- Braced wall panels must be constructed with wood structural panels having a minimum thickness of 7/16 inch with 8d common nails (0.131 x 2-1/2 inches) spaced 4 inches on center at panel ends and edges.
- When a hold-down post is located within a braced wall panel, 8d common nails spaced 4 inches on center are required for the sheathing attachment to the hold-down post.
- The amount of wall bracing is determined by IRC Table R703.7(2) rather than IRC Table R602.10.1.

- Cripple walls are not permitted.
- Interior braced wall lines are required to be supported on footings.
- Hold-down devices are required when masonry veneer is used in Seismic Design Categories D_0, D_1, and D_2. The capacity of the hold down is determined by IRC Table R703.7(2). This table lists capacities when the ends of braced wall panels are vertically aligned between stories and when they are not aligned. The ends of all braced wall lines must have hold downs (Section R703.7, Exception 2).
- When braced wall panels are not vertically aligned, an additional hold down must be installed beneath all hold-down locations in the story above per IRC Figure R703.7(1).

Height limits are illustrated in Figure 4.19.

FIGURE 4.19

SUMMARY OF STONE AND MASONRY REQUIREMENTS OF R703 FOR SDC D_0, D_1 AND D_2

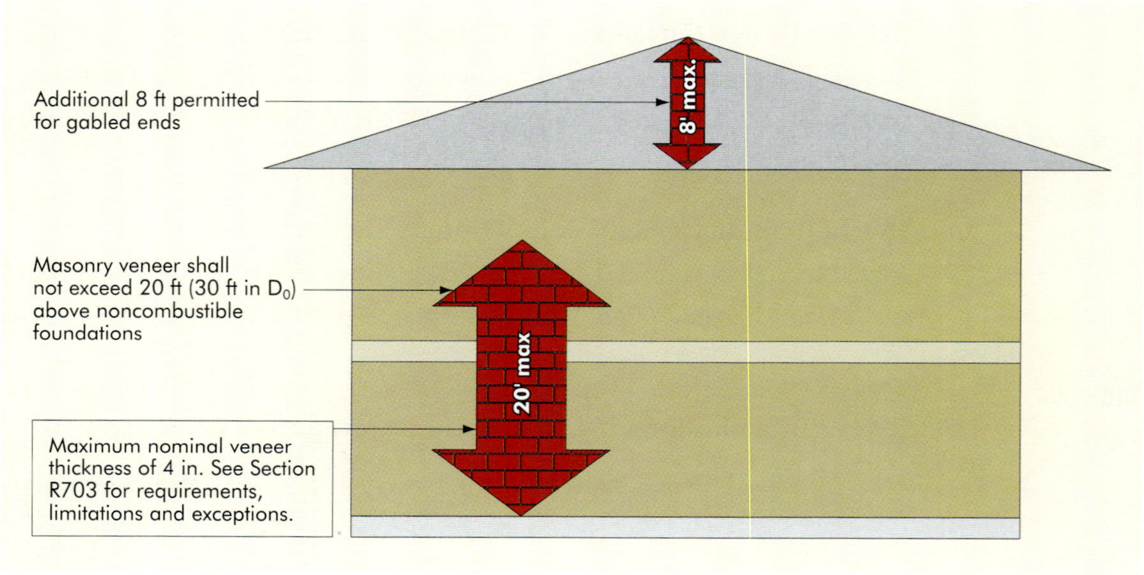

7) Snow loads

For all construction types, 70 psf is the maximum ground snow load permitted for buildings constructed per the IRC (R301.2.3). Buildings with a ground snow load greater than 70 psf must be designed in accordance with accepted engineering practice.

CHAPTER 5

Description of Bracing Methods

When it comes to selecting the bracing for a specific braced wall line in a residential structure, the builder is provided with a number of wall bracing options.

Overview

The eight isolated bracing methods

There are eight methods permitted in Sections R602.10.3 of the IRC. These are the traditional bracing methods, including let-in bracing, diagonal board bracing and six panel-type bracing methods, listed below:

1. Let-in diagonal brace
2. 5/8-inch diagonal wood boards
3. Wood structural panel (plywood or OSB)
4. 1/2-inch structural fiberboard
5. 1/2-inch interior gypsum wallboard or gypsum sheathing
6. Particleboard
7. Portland cement plaster on studs
8. Hardboard panel siding

Using these traditional methods, the minimum required length of a single braced wall panel segment ranges from 48 to 96 inches. These eight methods will be discussed in detail and illustrated later in this chapter. The methods are referred to as isolated bracing in this book because they may occur in isolated portions of the wall, as shown in Figure 5.1.

FIGURE 5.1

EXAMPLE OF "ISOLATED" METHOD 3 BRACING PANEL

Isolated bracing needs only to occur in isolated, specified locations.

Narrow alternate bracing with hold downs

In addition to the eight isolated bracing methods, there are two alternates described in Sections R602.10.6.1 and R602.10.6.2. These range in minimum lengths from 16 to 32 inches and may be substituted for *any* bracing method permitted in Sections R602.10.3 of the IRC. Both of these methods utilize prefabricated metal hold downs, in addition to anchor bolts, to connect the bracing section to the foundation below. For purposes of computing percent bracing to meet the requirements of Table R602.10.1 (described in more detail in Chapter 6), both alternate methods are recognized in the code to be equivalent to 48 inches of bracing. These are also referred to as isolated bracing in this book because they may occur in isolated portions of the wall, as shown in Figure 5.1.

HOLD DOWN – A prefabricated metal anchoring device that attaches the framing of a wall system to the structure below. Ultimately, the hold down load path must extend down into the foundation. The hold down prevents uplift of the studs and, thus, overturning of the wall segment.

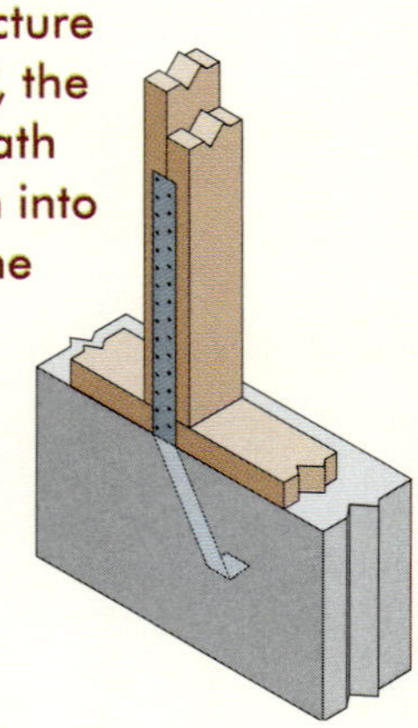

Continuous wood structural panel bracing

The continuous wood structural panel sheathing bracing method, described in Section R602.10.5 of the 2006 IRC, requires all exterior and interior braced wall lines, including areas above and below openings and gable ends, to be fully sheathed with a minimum of 5/16-inch wood structural panel sheathing, as shown in Figure 5.2.

FIGURE 5.2
EXAMPLES OF CONTINUOUS WOOD STRUCTURAL PANEL BRACED WALLS

In addition to continuously sheathing each wall, all corners must be constructed in accordance with Figure R602.10.5, as shown in Figure 5.3.

FIGURE 5.3

R602.10.5 CORNER DETAIL REQUIREMENT FOR ALL CONTINUOUS WOOD STRUCTURAL PANEL SHEATHING

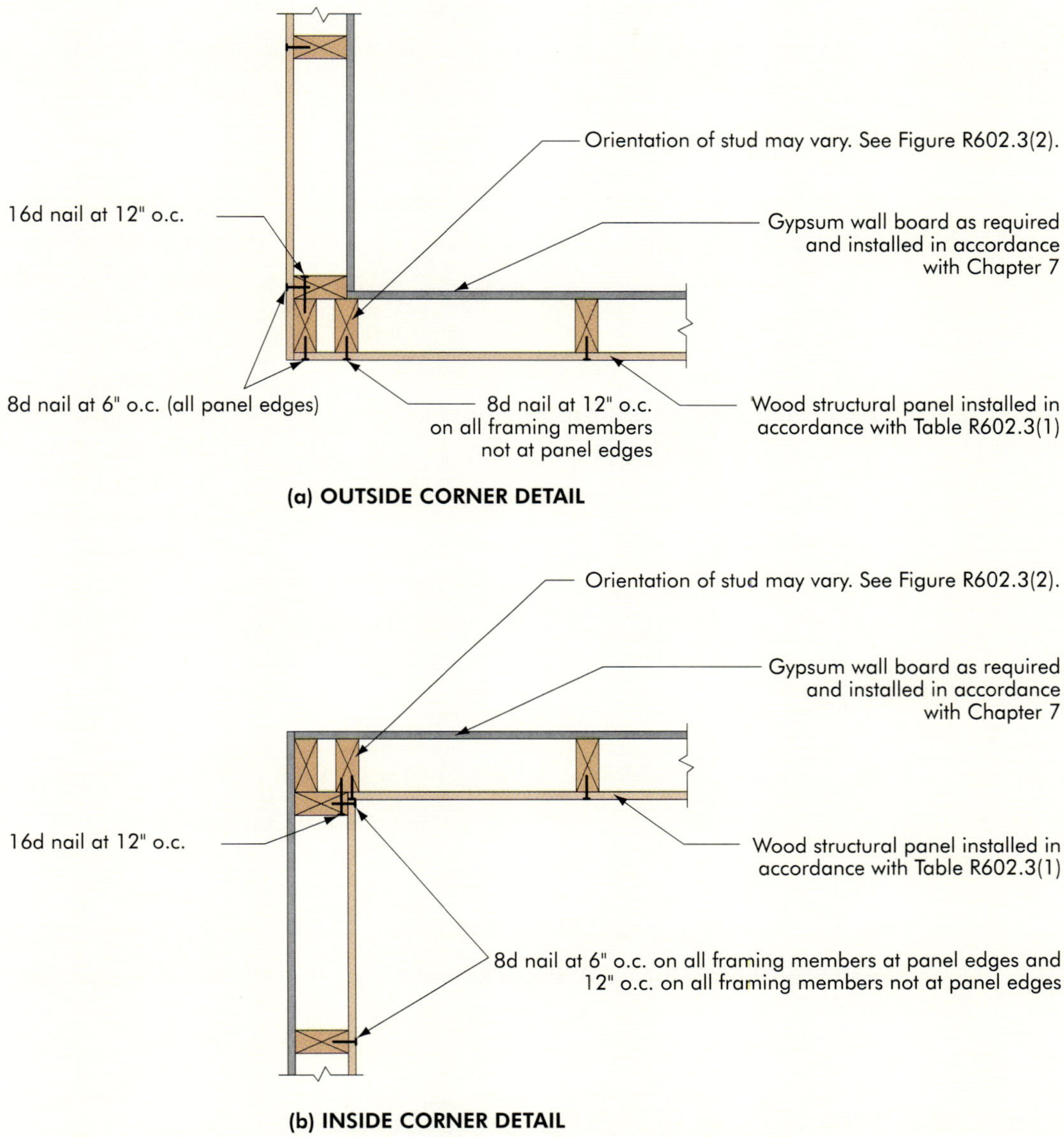

For SI: 1 inch = 25.4 mm.
Gypsum board nails omitted for clarity.

The purpose of the corner attachment is to connect the intersecting walls together to create a stronger "box-like" structure to perform better during high wind or seismic events. The corner detail requires a minimum of a single row of 16d nails at 12 inches on center. A double row of 16d nails located at 24 inches on center, framing member orientation permitting, is considered equivalent. It is important to note that the intent of Figure R602.10.5 is to provide the specified amount of nailing between the two studs, each on adjacent walls, to which the adjacent wall sheathing is attached.

Because of the added strength and stiffness of wood structural panels on all exterior surfaces (that are not window or door openings), narrower bracing panels may be used to give additional architectural latitude. Panels as narrow as 24 inches in length (4:1 aspect ratio) may be used as bracing for an 8-foot-high wall. Longer braced panels are required for taller walls. In addition, the minimum length of the bracing panel depends on the size of the opening next to the bracing panel. The relationship of bracing panel length relative to wall and adjacent opening height is explained later in this chapter.

ASPECT RATIO = The ratio of the height of a bracing unit to its length. For example, a 24-inch-wide section in an 8-foot wall has an aspect ratio of 4:1 (8'/2'). Also known as height-to-length ratio.

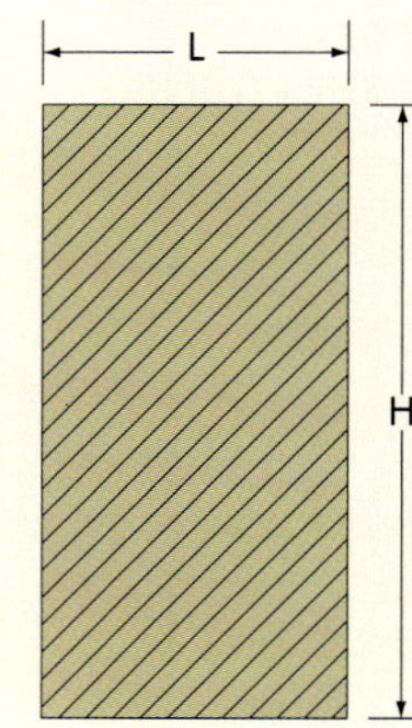

Panel lengths as narrow as 24 inches (4:1 aspect ratio) may be used adjacent to garage door openings with very light roof loads (Table R602.10.5, Footnote b).

Table R602.10.5, Footnote c, also permits the use of site-built portal frames *without hold downs* adjacent to garage doors with an aspect ratio of 6:1. This footnote permits site-built portal frames in single story and first of two-story applications in Seismic Design Categories (SDC) A–C.

Description of bracing methods

Table 5.1 shows all bracing methods and alternatives described in the IRC and discussed in this chapter. Note that the following figures show a permissible wall height of 8–12 feet. This is based on Section R301.3. While Chapter 3 of this publication covers this in more detail, Section R301.3, item 1 exception, permits the wall height to increase to 12 feet, provided the length of the required bracing in that wall is increased by multiplying by 1.2.

TABLE 5.1

Description of bracing methods

IRC Reference	Description	Figure	Minimum Length	Note
Method 1 R602.10.3	1" x 4" diagonal bracing	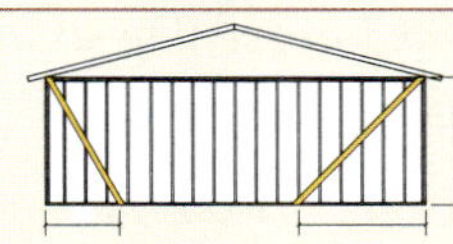	From 55" to 96" (60 to 45 degrees from horizontal)	8' wall = 4'-7" to 8' 9' wall = 5'-2" to 9' 10' wall = 5'-9" to 10'
Method 2 R602.10.3	Diagonal wood boards	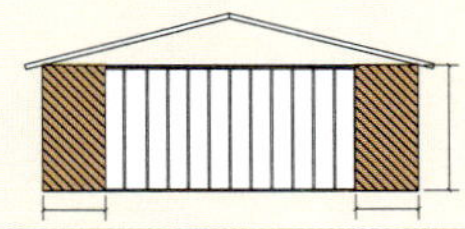	4'	Attach per Table R602.3(1)
Method 3 R602.10.3	Wood structural panel	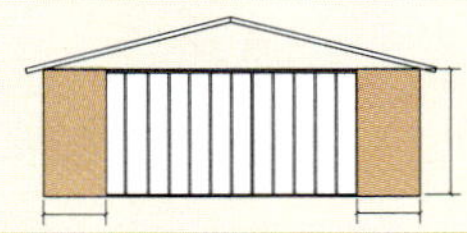	4'	Attach per Table R602.3(3)
Method 4 R602.10.3	Structural fiberboard	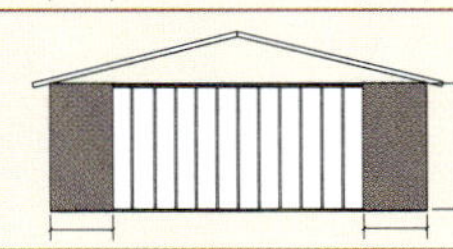	4'	Attach per Table R602.3(1)
Method 5 R602.10.3	Gypsum wallboard	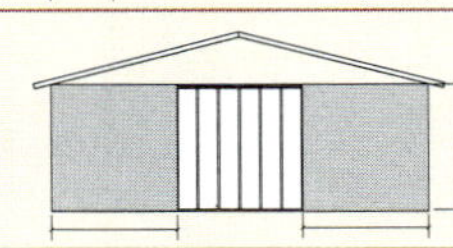	8' one side, 4' both sides	Attach per Table R602.3(1) for sheathing, Table R702.3.5 for interior gypsum board
Method 6 R602.10.3	Particleboard	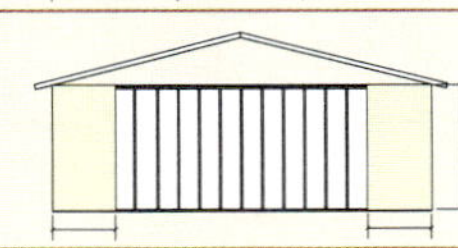	4'	Attach per Table R602.3(4)
Method 7 R602.10.3	Portland cement plaster on studs	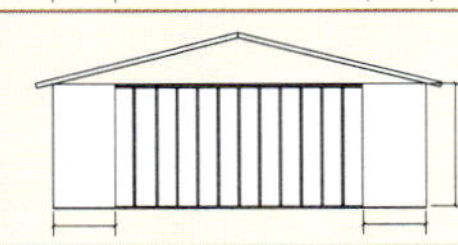	4'	Attach per Section 703.6
Method 8 R602.10.3	Hardboard panel siding	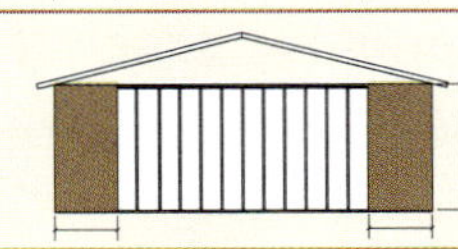	4'	Attach per Table 703.4
R602.10.6.1	Bracing panel with hold downs	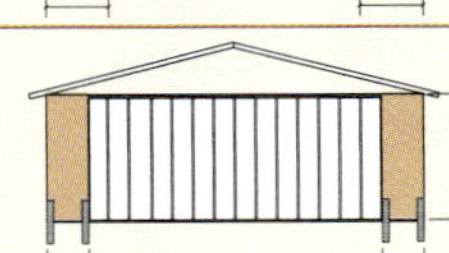	28" one side – no floor above 28" two sides – one floor above	With hold downs. Min. length varies 28" to 42"
R602.10.6.2	Portal frame with hold downs	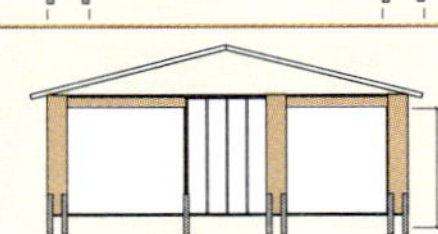	16" – no floor above 24" – one floor above	With hold downs

Continuous sheathing qualifying segments

IRC Reference	Description	Figure	Minimum Length	Note
R602.10.5	Wood structural panel, continuously sheathed	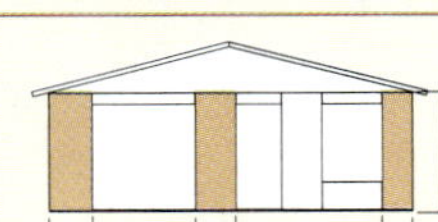	24" and up	All exterior surfaces sheathed except openings
Table R602.10.5, Footnote b	Wood structural panel, continuously sheathed	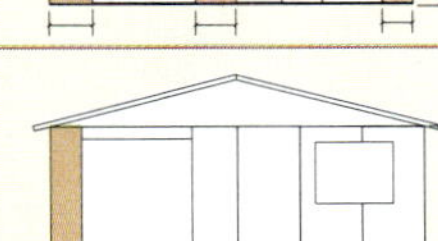	8' wall = 24" 9' wall = 27" 10' wall = 30"	No hold downs, adjacent to garage door opening, light roof, single story, 4:1 aspect ratio
Table R602.10.5, Footnote c	Portal frame without hold downs, continuously sheathed	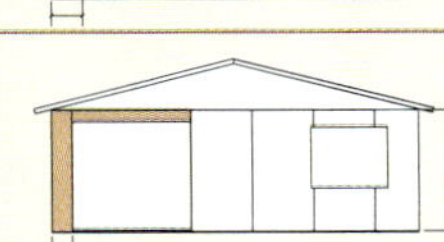	8' wall = 16" 9' wall = 18" 10' wall = 20"	No hold downs, adjacent to garage door opening, first of two-story, 6:1 aspect ratio

Bracing Method 1: (The following excerpt taken from Section R602.10.3)

> *1. Nominal 1- by 4-inch (25.4- by 102-mm) continuous diagonal braces let in to the top and bottom plates and the intervening studs or approved metal strap devices installed in accordance with the manufacturer's specifications. The let-in bracing shall be placed at an angle not more than 60 degrees (1.06 rad) or less than 45 degrees (0.79 rad) from the horizontal.* (See Figure 5.4)

FIGURE 5.4

METHOD 1 – DIAGONAL 1x4 LET-IN BRACING

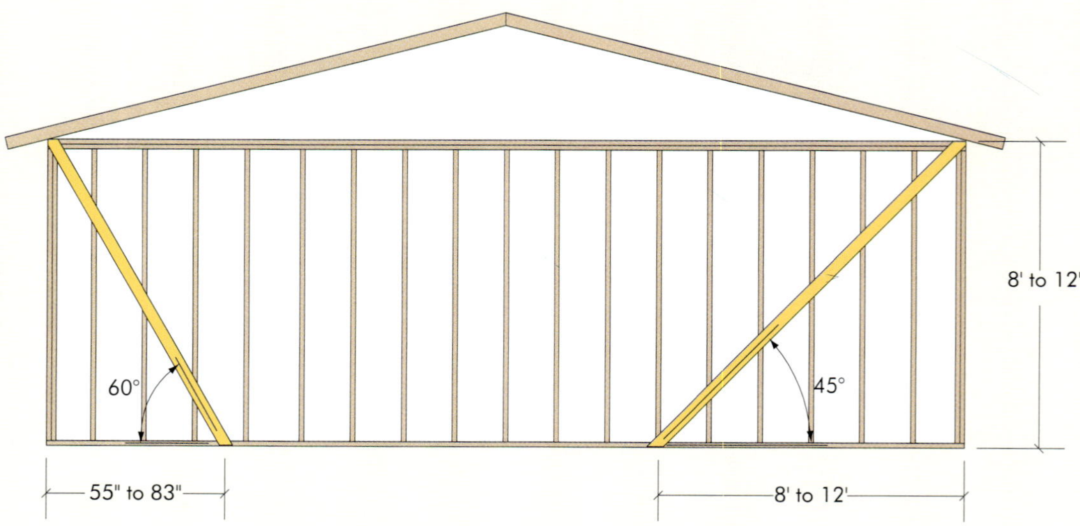

Fastening requirements for bracing Method 1

Excerpt from *Table R602.3(1) – Fastening Schedule for Structural Members*

Description of Building Element	Number and Type of Fastener[a,b]	Spacing of Fasteners
1" brace to each stud and plate, face nail	2-8d (2-1/2" x 0.113")	–
	2 staples, 1-3/4"	–

a. All nails are smooth-common, box or deformed shanks except where otherwise stated. Nails used for framing and sheathing connections shall have minimum average bending yield strengths as shown: 80 ksi for shank diameter of 0.192 inch (20d common nail), 90 ksi for shank diameter larger than 0.142 inch but not larger than 0.177 inch, and 100 ksi for shank diameters of 0.142 inch or less.

b. Staples are 16-gage wire and have a minimum 7/16-inch on diameter crown width.

Bracing Method 2: (The following excerpt taken from Section R602.10.3)

> 2. *Wood boards of 5/8-inch (15.9 mm) net minimum thickness applied diagonally on studs spaced a maximum of 24 inches (610 mm). Diagonal boards shall be attached to studs in accordance with Table R602.3(1).* (See Figure 5.5)

FIGURE 5.5

METHOD 2 – DIAGONAL WOOD BOARD BRACING

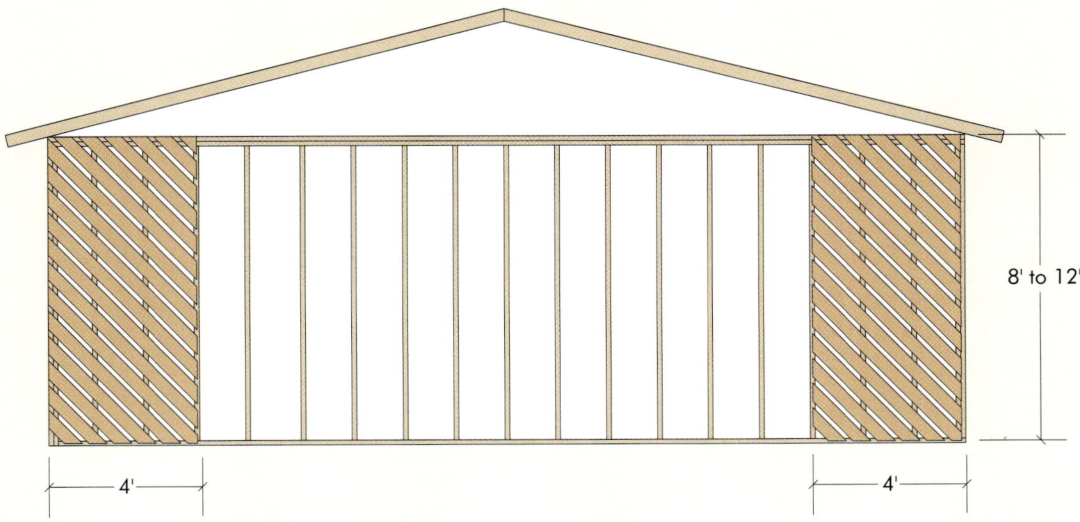

Note: Spacing between boards exagerated for clarity.

Fastening requirements for bracing Method 2

Excerpt from *Table R602.3(1) – Fastening Schedule for Structural Members*

Description of Building Element	Number and Type of Fastener[a,b]	Spacing of Fasteners
1" x 6" sheathing to each bearing, face nail	2-8d (2-1/2" x 0.113")	–
	2 staples, 1-3/4"	–
1" x 8" sheathing to each bearing, face nail	2-8d (2-1/2" x 0.113")	–
	3 staples, 1-3/4"	–
Wider than 1" x 8" sheathing to each bearing, face nail	3-8d (2-1/2" x 0.113")	–
	4 staples, 1-3/4"	–

a. All nails are smooth-common, box or deformed shanks except where otherwise stated. Nails used for framing and sheathing connections shall have minimum average bending yield strengths as shown: 80 ksi for shank diameter of 0.192 inch (20d common nail), 90 ksi for shank diameter larger than 0.142 inch but not larger than 0.177 inch, and 100 ksi for shank diameters of 0.142 inch or less.

b. Staples are 16-gage wire and have a minimum 7/16-inch on diameter crown width.

Bracing Method 3: (The following excerpt taken from Section R602.10.3)

3. Wood structural panel sheathing with a thickness not less than 5/16 inch (7.9 mm) for 16-inch (406 mm) stud spacing and not less than 3/8 inch (9.5 mm) for 24-inch (610 mm) stud spacing. Wood structural panels shall be installed in accordance with Table R602.3(3). (See Figure 5.6)

FIGURE 5.6

METHOD 3 – WOOD STRUCTURAL PANEL SHEATHING BRACING

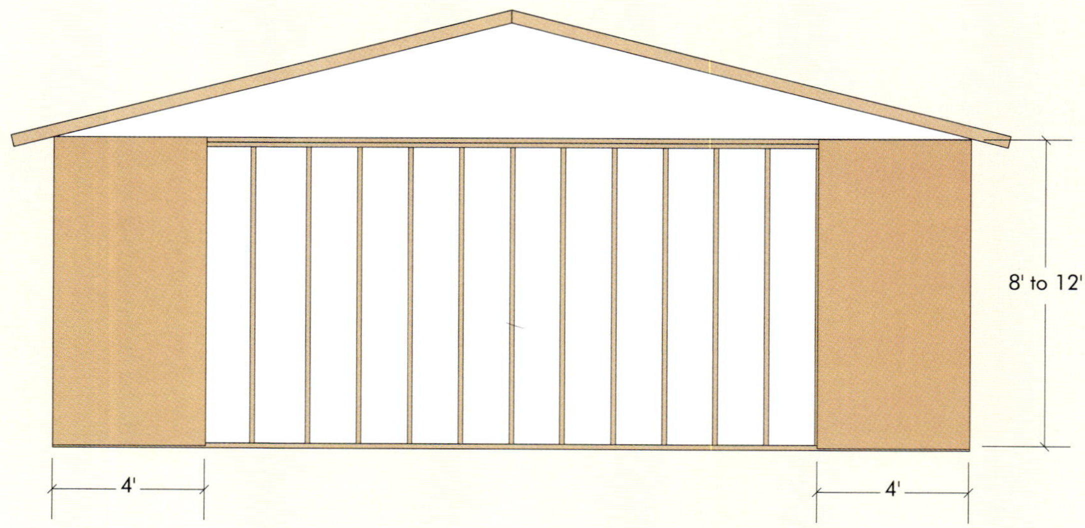

Fastening requirements for bracing Method 3

Excerpt from *Table R602.3(1) – continued – Fastening Schedule for Structural Members*

Description of Building Materials	Description of Fastener[c]	Spacing of Fasteners: Edges (inches)	Spacing of Fasteners: Intermediate Supports[c] (inches)
Wood structural panel wall sheathing to framing			
5/16"–1/2"	6d common (2" x 0.113") nail (subfloor and wall)	6	12[g]

c. Nails shall be spaced at not more than 6 inches on center at all supports where spans are 48 inches or greater.

g. For regions having basic wind speed of 100 mph or less, nails for attaching wood structural panel roof sheathing to gable end wall framing shall be spaced 6 inches on center. When basic wind speed is greater than 100 mph, nails for attaching panel roof sheathing to intermediate supports shall be spaced 6 inches on center for minimum 48-inch distance from ridges, eaves and gable end walls; and 4 inches on center to gable end wall framing.

Bracing Method 4: (The following excerpt taken from Section R602.10.3)

> *4. One-half-inch (12.7 mm) or 25/32-inch-thick (19.8 mm) structural fiberboard sheathing applied vertically or horizontally on studs spaced a maximum of 16 inches (406 mm) on center. Structural fiberboard sheathing shall be installed in accordance with Table R602.3(1).* (See Figure 5.7)

FIGURE 5.7

METHOD 4 – STRUCTURAL FIBERBOARD BRACING

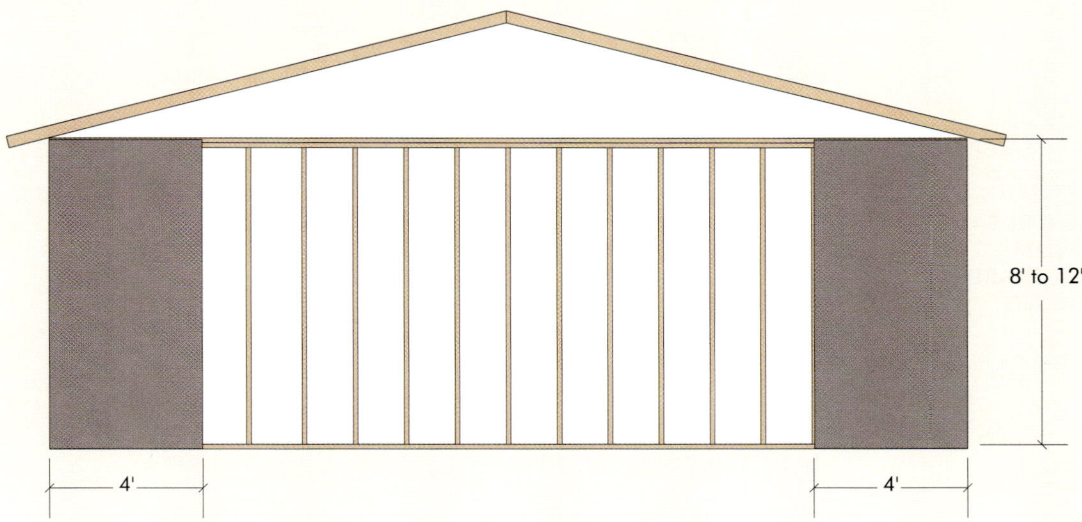

Fastening requirements for bracing Method 4

Excerpt from *Table R602.3(1) – continued – Fastening Schedule for Structural Members*

Description of Building Materials	Description of Fastener[b,c,e]	Spacing of Fasteners: Edges (inches)	Spacing of Fasteners: Intermediate Supports[c,e] (inches)
	Other wall Sheathing[h]		
1/2" structural cellulosic fiberboard sheathing	1-1/2" galvanized roofing nail; 8d common (2-1/2" x 0.131") nail; staple 16 ga., 1-1/2" long	3	6
25/32" structural cellulosic fiberboard sheathing	1-3/4" galvanized roofing nail; 8d common (2-1/2" x 0.131") nail; staple 16 ga., 1-3/4" long	3	6

b. Staples are 16-gage wire and have a minimum 7/16-inch on diameter crown width.

c. Nails shall be spaced at not more than 6 inches on center at all supports where spans are 48 inches or greater.

e. Spacing of fasteners not included in this table shall be based on Table R602.3(2).

h. Gypsum sheathing shall conform to ASTM C 79 and shall be installed in accordance with GA 253. Fiberboard sheathing shall conform to ASTM C 208.

Bracing Method 5: (The following excerpt taken from Section R602.10.3)

> *5. Gypsum board with minimum 1/2-inch (12.7 mm) thickness placed on studs spaced a maximum of 24 inches (610 mm) on center and fastened at 7 inches (178 mm) on center with the size nails specified in Table R602.3(1) for sheathing and Table R702.3.5 for interior gypsum board.* (See Figure 5.8)

Note that the two tables in Method 5 are for different gypsum products. The first is for typical gypsum wallboard and the second is for gypsum sheathing. Gypsum sheathing is a product meant for protected exterior wall applications. It is formulated to have greater resistance to moisture and is often used on structures to resist the spread of fire from one structure to the next.

FIGURE 5.8

METHOD 5 – GYPSUM WALLBOARD BRACING

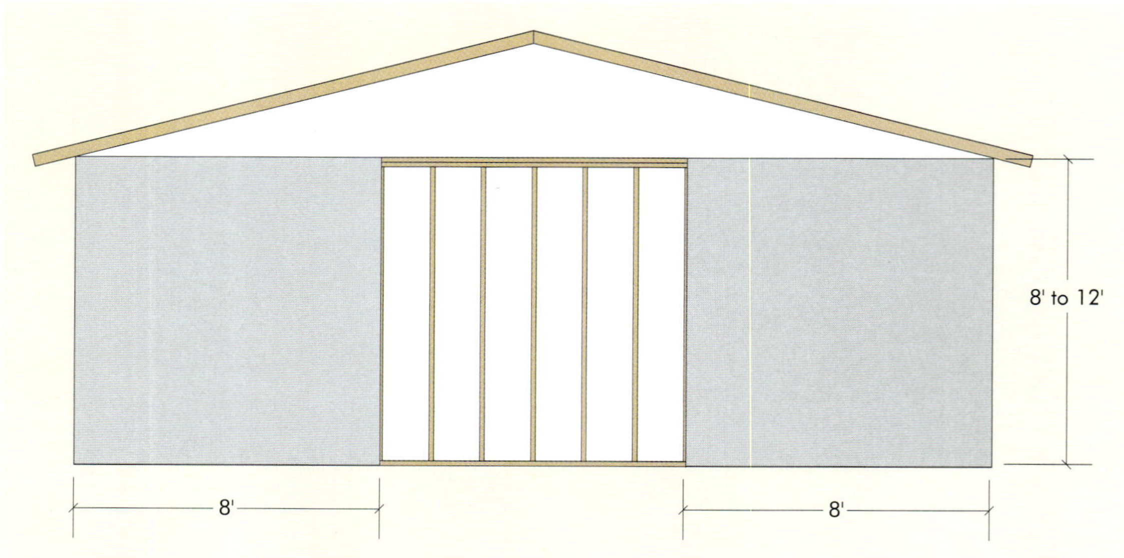

Note: the widths indicated above are for single-sided gypsum bracing panels. Section R602.10.4 permits a minimum length of 4 feet for gypsum placed on both faces of the wall.

Fastener requirements for bracing Method 5 – gypsum sheathing

Excerpt from *Table R602.3(1) – continued – Fastening Schedule for Structural Members*

Description of Building Materials	Description of Fastener[b]	Spacing of Fasteners: Edges (inches)	Spacing of Fasteners: Intermediate Supports (inches)
Other wall sheathing[h]			
1/2" gypsum sheathing[d]	1-1/2" galvanized roofing nail; 6d common nail (2 x 0.113"); staple galvanized, 1-1/2" long; 1-1/4" screws, Type W or S	Not applicable for bracing	
5/8" gypsum sheathing[d]	1-3/4" galvanized roofing nail; 8d common nail (2-1/2" x 0.131"); staple galvanized, 1-5/8" long; 1-5/8" screws, Type W or S	Not applicable for bracing	

b. Staples are 16-gage wire and have a minimum 7/16-inch on diameter crown width.

d. Four-foot-by-8-foot or 4-foot-by-9-foot.

h. Gypsum sheathing shall conform to ASTM C 79 and shall be installed in accordance with GA 253. Fiberboard sheathing shall conform to ASTM C 208.

Fastener requirements for bracing Method 5 – interior gypsum board

Excerpt from *Table R702.3.5 – Minimum Thickness and Application of Gypsum Board*

Thickness of Gypsum Board (inches)	Application	Orientation of Gypsum Board to Framing	Maximum Spacing of Framing Members (inches)	Maximum Spacing of Fasteners (inches): Nails	Maximum Spacing of Fasteners (inches): Screws[b]	Size of Nails for Application to Wood Framing
Application without adhesive						
1/2	Wall	Either direction	–	Not applicable for bracing		13 gage, 1-3/8" long, 19/64" head; 0.098" diameter, 1-1/4" long, annular-ringed; 5d cooler nail, 0.086" diameter, 1-5/8" long, 15/64" head; or gypsum board nail, 0.086" diameter, 1-5/8" long, 9/32" head
	Wall	Either direction	–	Not applicable for bracing		
5/8	Wall	Either direction	–	Not applicable for bracing		13 gage, 1-5/8" long, 19/64" head; 0.098" diameter, 1-3/8" long, annular-ringed; 6d cooler nail, 0.092" diameter, 1-7/8" long, 1/4" head; or gypsum board nail, 0.0915" diameter, 1-7/8" long, 19/64" head
	Wall	Either direction	–	Not applicable for bracing		

b. Screws shall be Type S or W per ASTM C 1002 and shall be sufficiently long to penetrate wood framing not less than 5/8 inch and metal framing not less than 3/8 inch

Bracing Method 6: (The following excerpt taken from Section R602.10.3)

6. Particleboard wall sheathing panels installed in accordance with Table R602.3(4). (See Figure 5.9)

FIGURE 5.9

METHOD 6 – PARTICLEBOARD WALL SHEATHING BRACING

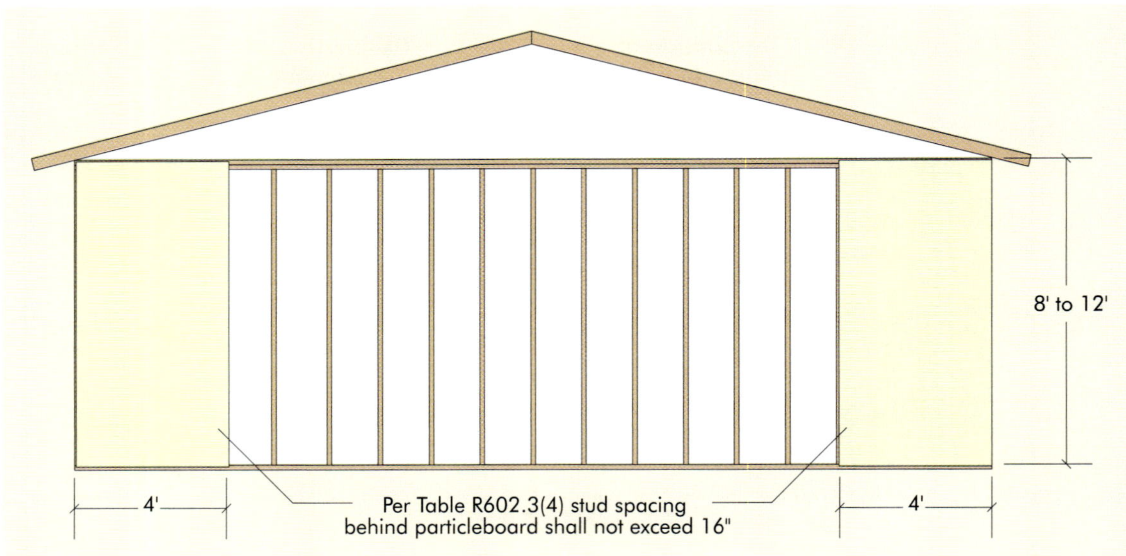

Fastening requirements for bracing Method 6

Excerpt from *Table R602.3(1) – continued – Fastening Schedule for Structural Members*

Description of Building Materials	Description of Fastener[c]	Spacing of Fasteners: Edges (inches)	Spacing of Fasteners: Intermediate Supports[c] (inches)
5/16"–1/2"	6d common nail (2" x 0.113") (subfloor and wall)	6	12[g]

c. Nails shall be spaced at not more than 6 inches on center at all supports where spans are 48 inches or greater.

g. For regions having basic wind speed of 100 mph or less, nails for attaching wood structural panel roof sheathing to gable end wall framing shall be spaced 6 inches on center. When basic wind speed is greater than 100 mph, nails for attaching panel roof sheathing to intermediate supports shall be spaced 6 inches on center for minimum 48-inch distance from ridges, eaves and gable end walls; and 4 inches on center to gable end wall framing.

Bracing Method 7: (The following excerpts taken from Section R602.10.3)

7. Portland cement plaster on studs spaced a maximum of 16 inches (406 mm) on center and installed in accordance with Section R703.6. (See Figure 5.10)

FIGURE 5.10
PORTLAND CEMENT PLASTER BRACING

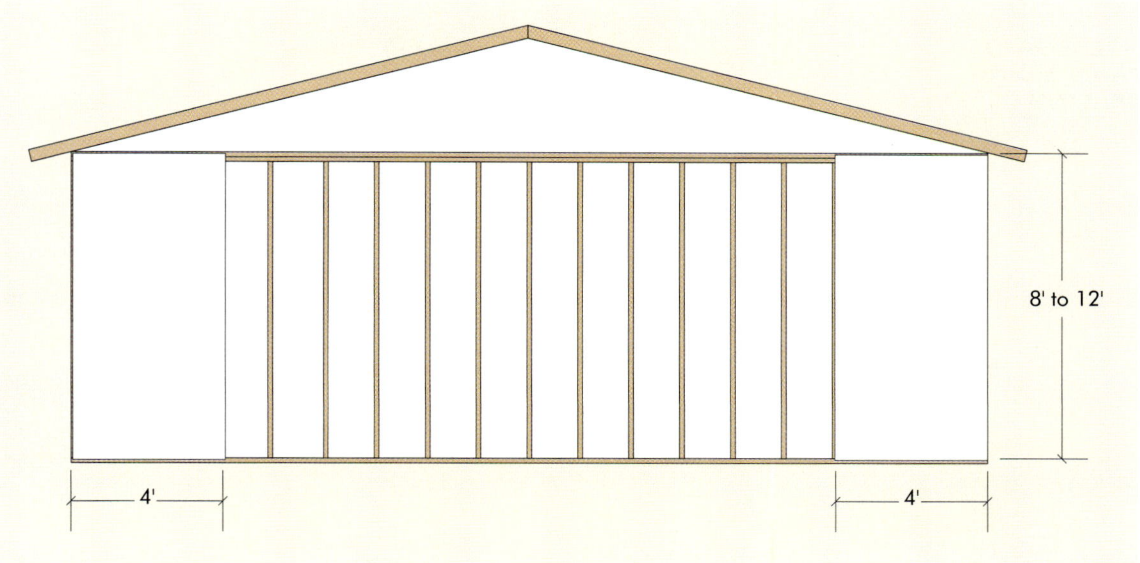

Fastening requirements for bracing Method 7 – portland cement plaster

R703.6 Exterior plaster. *Installation of these materials shall be in compliance with ASTM C 926 and ASTM C 1063 and the provisions of this code.*

R703.6.1 Lath. *All lath and lath attachment shall be of corrosion-resistant materials. Expanded metal or woven wire lath shall be attached with 1-1/2-inch-long (38 mm), 11-gage nails having a 7/16-inch (11.1 mm) head, or 7/8-inch-long (22.2 mm), 16-gage staples, spaced at no more than 6 inches (152 mm), or as otherwise approved.*

R703.6.2 Plaster. *Plastering with portland cement plaster shall not be less than three coats when applied to metal lath or wire lath and shall be not less than two coats over masonry, concrete, pressure-preservative-treated wood or decay-resistant wood as specified in Section R319.1 or gypsum backing. If the plaster surface is completely covered by veneer or other facing material, or is completely concealed, plaster application need be only two coats, provided the total thickness is as set forth in Table R702.1(1).*

Bracing Method 8: (The following excerpt taken from Section R602.10.3)

8. Hardboard panel siding when installed in accordance with Table R703.4. (See Figure 5.11)

FIGURE 5.11
HARDBOARD PANEL SIDING BRACING

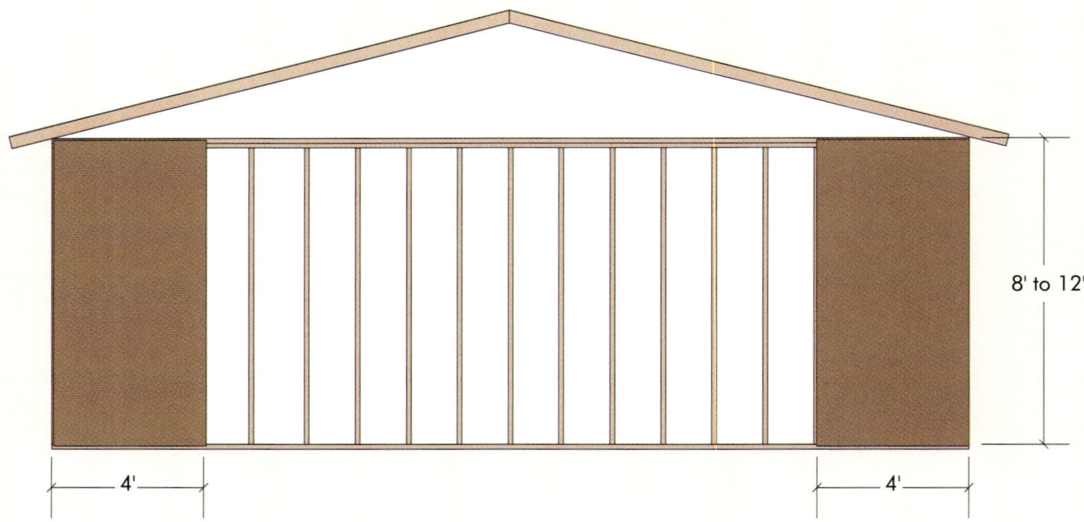

Fastening requirements for bracing Method 8

Excerpt from *Table R703.4 – Weather-resistant Siding Attachment and Minimum Thickness*

				Type of Support for the Siding Material and Fasteners[b,c,d]					
Siding Materials	**Nominal Thickness (inches)**	**Joint Treatment**	**Water-resistive Barrier Required**	**Wood or Wood Structural Panel Sheathing**	**Fiber-Board Sheathing into Stud**	**Gypsum Sheathing into Stud**	**Foam Plastic Sheathing into Stud**	**Direct to Stud**	**Number or Spacing of Fasteners**
Hardboard[k] panel siding – vertical	7/16	–	Yes	Note n	Note n	Note n	Note n	Note n	6" panel edges, 12" int. sup.[o]

b. Nail is a general description and shall be T-head, modified round head, or round head with smooth or deformed shanks.

c. Staples shall have a minimum crown width of 7/16-inch outside diameter and be manufactured of minimum 16-gage wire.

d. Nails or staples shall be aluminum, galvanized, or rust-preventative coated and shall be driven into the studs for fiberboard or gypsum backing.

k. Hardboard siding shall comply with AHA A135.6.

n. Minimum shank diameter of 0.092 inch, minimum head diameter of 0.225 inch, and nail length must accommodate sheathing and penetrate framing 1-1/2 inches.

o. **When used to resist shear forces, the spacing must be 4 inches at panel edges and 8 inches on interior supports.** *(Bold added for emphasis.)*

Alternate braced wall panels with hold downs (R602.10.6). Alternate braced wall panels are typically used when bracing is needed in a wall area that is not long enough to accommodate a 48-inch braced wall panel. Alternate braced wall panels range from 16 to 42 inches in length. They are limited to first stories and require the installation of hold downs. The minimum panel length depends on the construction method used and if the wall will be supporting a second floor or roof only. Table 5.2 summarizes the minimum lengths, construction methods and story applications of alternate braced wall panels.

TABLE 5.2

Alternate braced wall panels with hold downs

Minimum Length (inches)	Construction Requirements	1 Story	1st of 2 Stories
28–42	Sheathing one side (R602.10.6.1)	X	
28–42	Sheathing both sides (R602.10.6.1)		X
16	Portal frame (R602.10.6.2)	X	
24	Portal frame (R602.10.6.2)		X

R602.10.6.1 Alternate braced wall panels. *Alternate braced wall lines constructed in accordance with one of the following provisions shall be permitted to replace each 4 feet (1219 mm) of braced wall panel as required by Section R602.10.4. The maximum height and minimum width of each panel shall be in accordance with Table R602.10.6:*

1. In one-story buildings, each panel shall be sheathed on one face with 3/8-inch minimum thickness (9.5 mm) wood structural panel sheathing nailed with 8d common or galvanized box nails in accordance with Table R602.3(1) and blocked at all wood structural panel sheathing edges. Two anchor bolts installed in accordance with Figure R403.1(1) shall be provided in each panel. Anchor bolts shall be placed at panel quarter points. Each panel end stud shall have a tie-down device fastened to the foundation, capable of providing an uplift capacity in accordance with Table R602.10.6. The tie-down device shall be installed in accordance with the manufacturer's recommendations. The panels shall be supported directly on a foundation or on floor framing supported directly on a foundation which is continuous across the entire length of the braced wall line. This foundation shall be reinforced with not less than one No. 4 bar top and bottom. When the continuous foundation is required to have a depth greater than 12 inches (305 mm), a minimum 12-inch-by-12-inch (305 mm by 305 mm) continuous footing or turned down slab edge is permitted at door openings in the braced wall line. This continuous footing or turned down slab edge shall be reinforced with not less than one No. 4 bar tip and bottom. This reinforcement shall be lapped 15 inches (381 mm) with the reinforcement required in the continuous foundation located directly under the braced wall line.

2. In the first story of two-story buildings, each braced wall panel shall be in accordance with Item 1 above, except that the wood structural panel sheathing shall be provided on both faces, sheathing edge nailing spacing shall not exceed four inches on center, at least three anchor bolts shall be placed at one-fifth points. (See Figures 5.12 and 5.13)

TABLE 5.3

TABLE R602.10.6
MINIMUM WIDTHS AND TIE-DOWN FORCES OF ALTERNATE BRACED WALL PANELS

Seismic Design Category and Windspeed		Height of Braced Wall Panel				
		8 ft	**9 ft**	**10 ft**	**11 ft**	**12 ft**
SDC A, B, and C Windspeed < 110 mph	Sheathed Length	2'-4"	2'-8"	2'-8"	3'-2"	3'-6"
	R602.10.6.1, Item 1 Tie-down Force (lbs)	1800	1800	1800	2000	2200
	R602.10.6.1, Item 2 Tie-down Force (lbs)	3000	3000	3000	3300	3600
SDC D_0, D_1 and D_2 Windspeed < 110 mph	Sheathed Length	2'-8"	2'-8"	2'-8"	NP[a]	NP[a]
	R602.10.6.1, Item 1 Tie-down Force (lbs)	1800	1800	1800	NP[a]	NP[a]
	R602.10.6.1, Item 2 Tie-down Force (lbs)	3000	3000	3000	NP[a]	NP[a]

For SI: 1 inch = 25.4 mm, 1 foot = 305 mm.
a. NP = Not Permitted. Maximum height of 10 feet.

FIGURE 5.12
BRACED WALL PANEL WITH HOLD DOWNS ALTERNATIVE

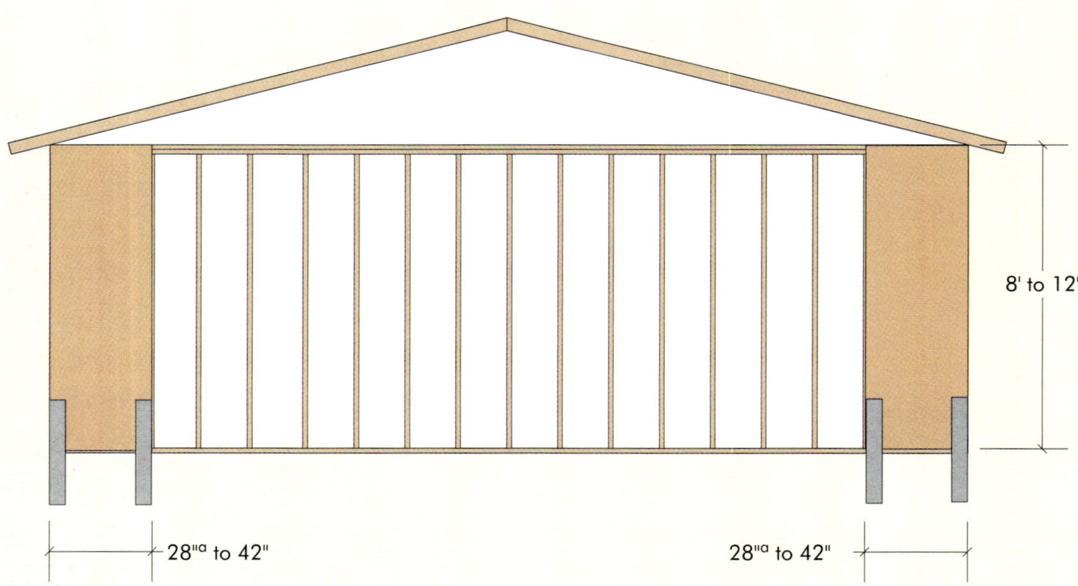

a. SDC D_0–D_2 limited to 10-ft max. height and 32-in. minimum length.

FIGURE 5.13

DETAILS OF BRACED WALL PANEL WITH HOLD DOWNS ALTERNATIVE (DETAIL FOR SECTION R602.10.6.1)

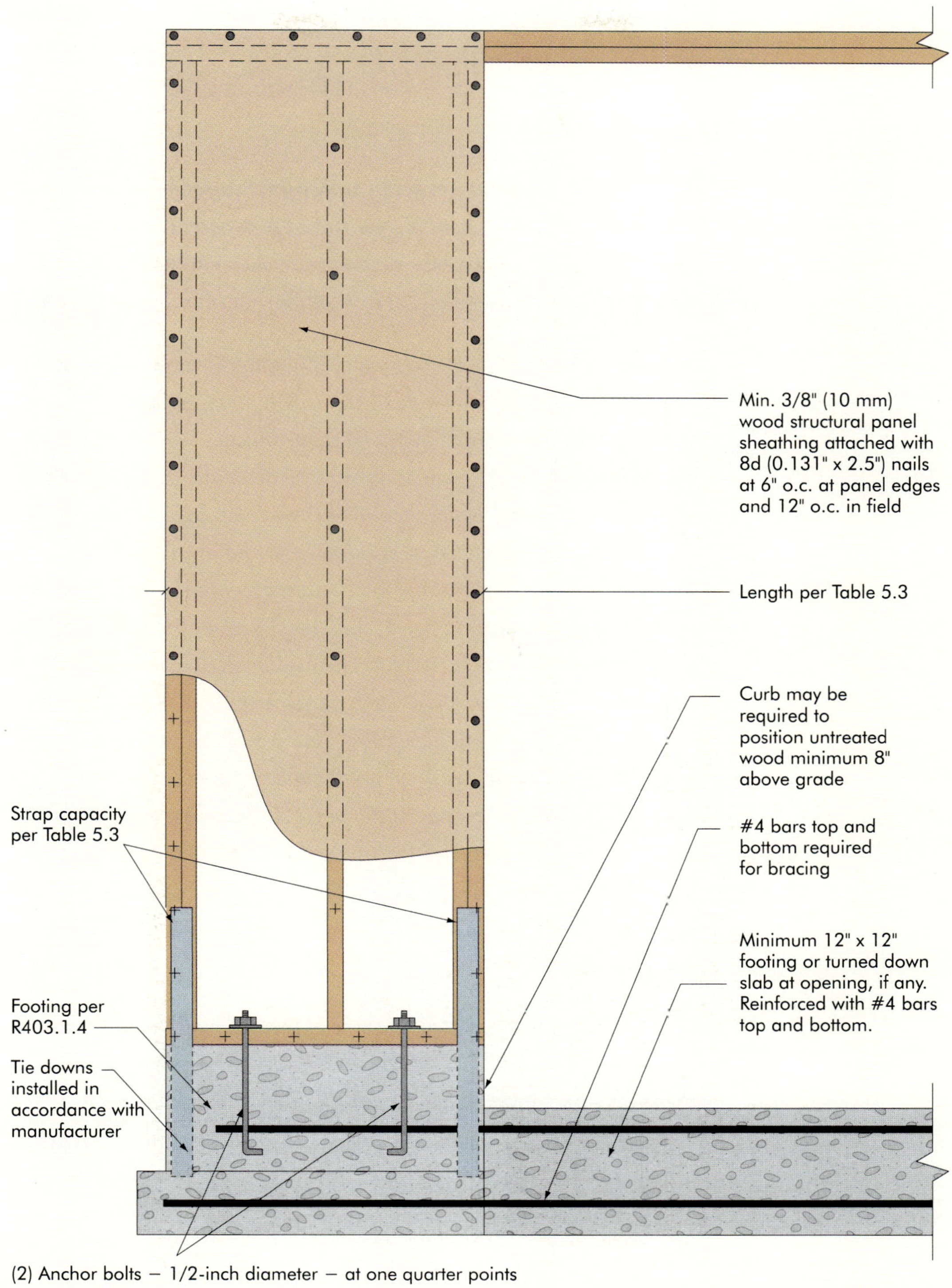

Wood structural panel portal frame with hold downs alternate method (R602.10.6.2). This alternate method prescribes a partially moment-resisting frame (a portal frame) to resist applied lateral loads. This is achieved by the required attachments at both the base of the wall and at the wall and header intersection. This method is permitted in a one-story building or on the first story of a two-story building only.

> ***R602.10.6.2 Alternate bracing wall panel adjacent to a door or window opening.*** *Alternate braced wall panels constructed in accordance with one of the following provisions are also permitted to replace each 4 feet (1219 mm) of braced wall panel as required by Section R602.10.4 for use adjacent to a window or door opening with a full-length header:*
>
> *1. In one-story buildings, each panel shall have a length of not less than 16 inches (406 mm) and a height of not more than 10 feet (3048 mm). Each panel shall be sheathed on one face with a single layer of 3/8-inch-minimum-thickness (9.5 mm) wood structural panel sheathing nailed with 8d common or galvanized box nails in accordance with Figure R602.10.6.2. The wood structural panel sheathing shall extend up over the solid sawn or glued-laminated header and shall be nailed in accordance with Figure R602.10.6.2. A built-up header consisting of at least two 2 x 12s and fastened in accordance with Table R602.3(1) shall be permitted to be used. A spacer, if used, shall be placed on the side of the built-up beam opposite the wood structural panel sheathing. The header shall extend between the inside faces of the first full-length outer studs of each panel. The clear span of the header between the inner studs of each panel shall be not less than 6 feet (1829 mm) and not more than 18 feet (5486 mm) in length. A strap with an uplift capacity of not less than 1000 pounds (4448 N) shall fasten the header to the side of the inner studs opposite the sheathing. One anchor bolt not less than 5/8-inch-diameter (16 mm) and installed in accordance with Section R403.1.6 shall be provided in the center of each sill plate. The studs at each end of the panel shall have a tie-down device fastened to the foundation with an uplift capacity of not less than 4,200 pounds (18683 N).*
>
> *Where a panel is located on one side of the opening, the header shall extend between the inside face of the first full-length stud of the panel and the bearing studs at the other end of the opening. A strap with an uplift capacity of not less than 1000 pounds (4448 N) shall fasten the header to the bearing studs. The bearing studs shall also have a tie-down device fastened to the foundation with an uplift capacity of not less than 1000 pounds (4448 N).*
>
> *The tie-down devices shall be an embedded-strap type, installed in accordance with the manufacturer's recommendations. The panels shall be supported directly on a foundation which is continuous across the entire length of the braced wall line. The foundation shall be reinforced with not less than one No. 4 bar top and bottom.*

Where the continuous foundation is required to have a depth greater than 12 inches (305 mm), a minimum 12-inch-by-12-inch (305 mm by 305 mm) continuous footing or turned down slab edge is permitted at door openings in the braced wall line. This continuous footing or turned down slab edge shall be reinforced with not less than one No. 4 bar top and bottom. This reinforcement shall be lapped not less than 15 inches (381 mm) with the reinforcement required in the continuous foundation located directly under the braced wall line. (See Figures 5.14 and 5.15)

2. In the first story of two-story buildings, each wall panel shall be braced in accordance with Item 1 above, except that each panel shall have a length of not less than 24 inches (610 mm). (See Figure 5.15 for details)

FIGURE 5.14
WOOD STRUCTURAL PANEL PORTAL FRAME WITH HOLD DOWNS ALTERNATE
R602.10.6.2

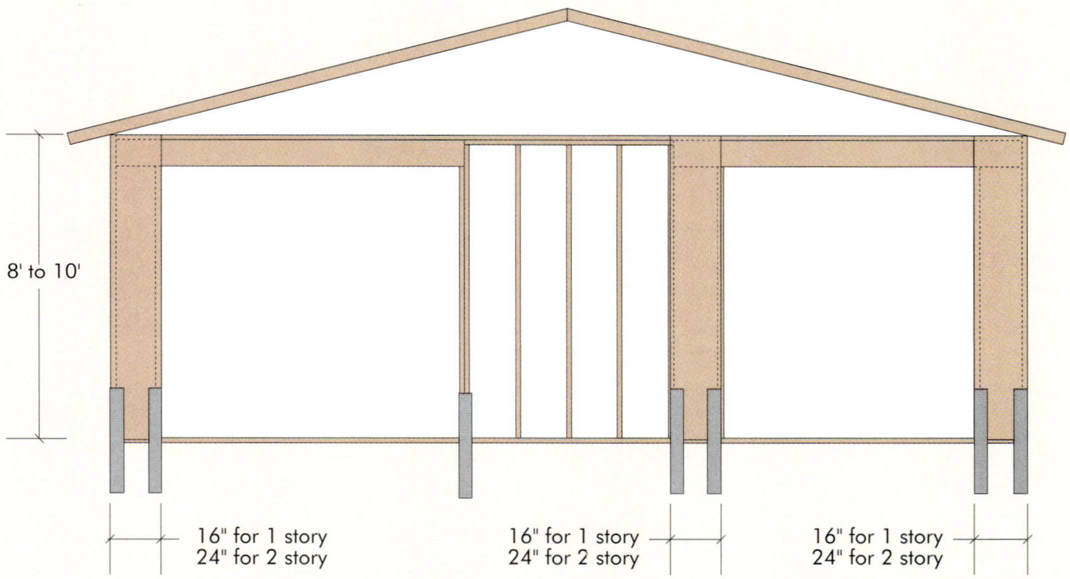

Note: Figure 5.14 shows a single and double portal frame as an illustration of its permitted configurations. The IRC permits the area between the portal frames with hold downs to be sheathed with any material as long as it complies with wall cladding requirements.

FIGURE 5.15

DETAILS OF PORTAL FRAME WITH HOLD DOWNS ALTERNATE – *R602.10.6.2*

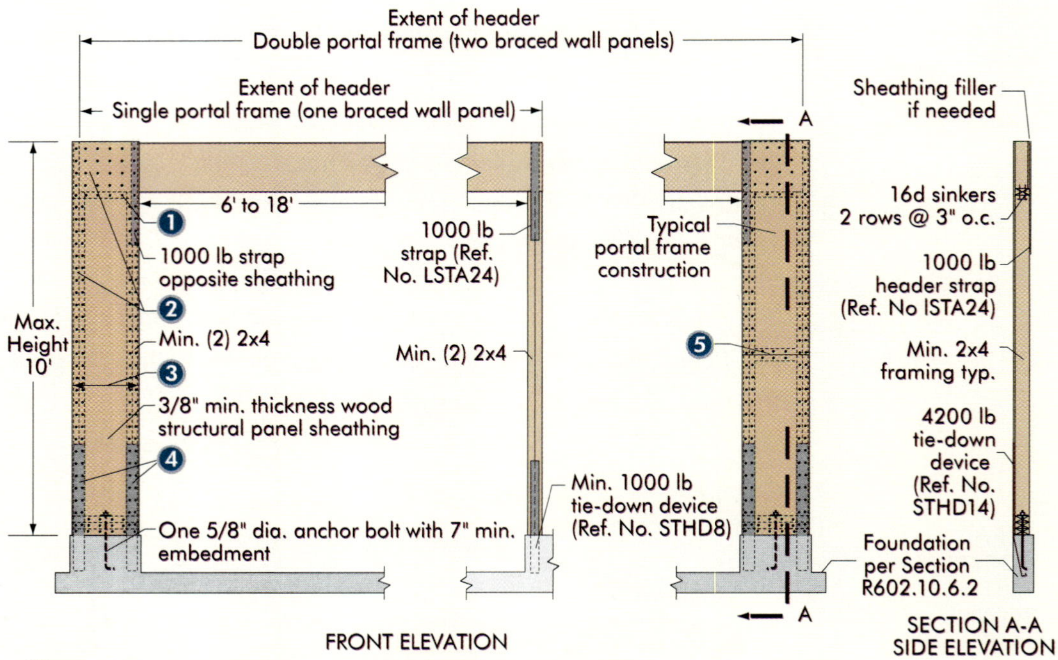

1. Fasten top plate to header with two rows of 16d sinker. Nails at 3" o.c. typ.
2. Fasten sheathing to header with 8d common or galvanized box nails in 3" grid pattern as shown and 3" o.c. in all framing (studs, blocking, and sills) typ.
3. For bracing: min. width = 16" for one story, min. width = 24" for use in the first of two stories.
4. Min. 4200 lb strap type tie-down device (embedded into concrete and nailed into framing). Installed per manufacturer. (Ref. No. STHD14)
5. For a panel splice (if needed), panel edges shall be blocked, and occur within 24" of mid-height. One row of typ. sheathing-to-framing nailing is required. If 2x4 blocking is used, the 2x4s must be nailed together with three 16d sinkers.

Note: Figure R602.10.6.2 (shown in Figure 5.15) illustrates that the portal frame can be used at either one or both sides of an opening. Each single portal frame is equivalent to one braced wall panel and equal to 4 ft of bracing. A portal frame on both sides of an opening is a double portal frame and is equivalent to **two** braced wall panels, or 8 ft of bracing.

Continuous sheathing qualifying segments

R602.10.5 – Continuous wood structural panel sheathing:

> ***R602.10.5 Continuous wood structural panel sheathing.*** *When continuous wood structural panel sheathing is provided in accordance with Method 3 of R602.10.3 on all sheathable areas of all exterior walls, and interior braced wall lines, where required, including areas above and below openings, braced wall panel lengths shall be in accordance with Table R602.10.5. Wood structural panel sheathing shall be installed at corners in accordance with Figure R602.10.5. The bracing amounts in Table R602.10.1 for Method 3 shall be permitted to be multiplied by a factor of 0.9 for walls with a maximum opening height that does not exceed 85 percent of the wall height or a factor of 0.8 for walls with a maximum opening height that does not exceed 67 percent of the wall height.* (See Figure 5.16)

TABLE 5.4

TABLE R602.10.5
LENGTH REQUIREMENTS FOR BRACED WALL PANELS IN A CONTINUOUSLY SHEATHED WALL[a,b,c]

MINIMUM LENGTH OF BRACED WALL PANEL (INCHES) FOR WALL HEIGHT OF:			MAXIMUM OPENING HEIGHT NEXT TO THE QUALIFIED BRACING SEGMENT (% OF WALL HEIGHT)
8-foot wall	9-foot wall	10-foot wall	
48	54	60	100%
32	36	40	85%
24	27	30	65%

For SI: 1 inch = 25.4 mm, 1 foot = 305 mm, 1 pound per square foot = 0.0479 kN/m^2

a. Linear interpolation shall be permitted.

b. Full-height sheathed wall segments to either side of garage openings that support light frame roofs only, with roof covering dead loads of 3 psf or less shall be permitted to have a 4:1 aspect ratio.

c. Walls on either or both sides of openings in garages attached to fully sheathed dwellings shall be permitted to be built in accordance with Section R602.10.6.2 and Figure R602.10.6.2 except that a single bottom plate shall be permitted and two anchor bolts shall be placed at 1/3 points. In addition, tie-down devices shall not be required and the vertical wall segment shall have a maximum 6:1 height-to-width ratio (with height being measured from top of header to the bottom of the sill plate). This option shall be permitted for the first story of two-story applications in Seismic Design Categories A through C.

FIGURE 5.16
CONTINUOUS WOOD STRUCTURAL PANEL SHEATHING
R602.10.5

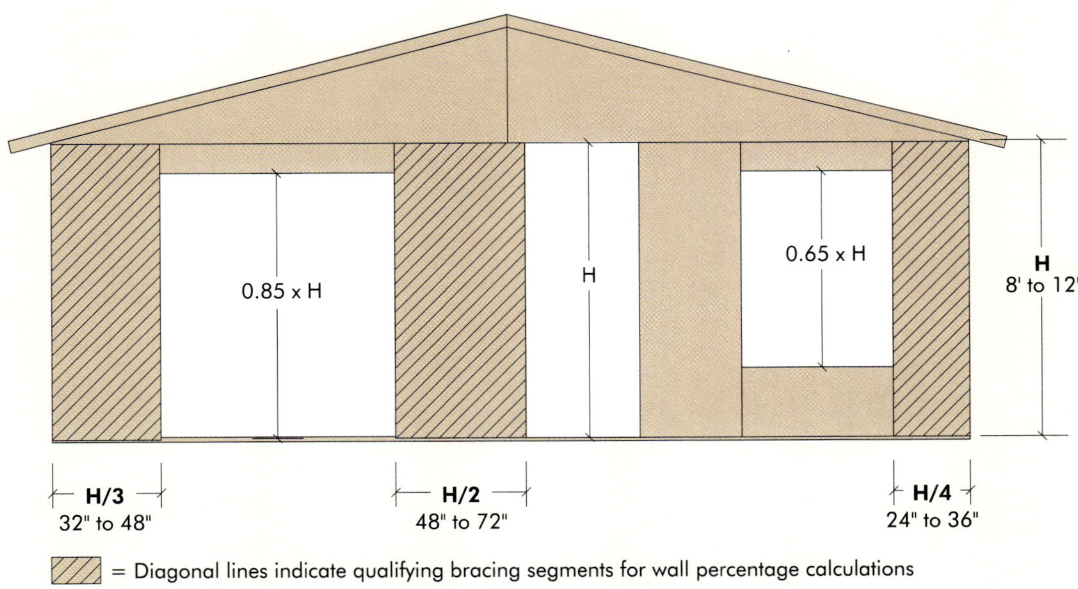

As mentioned previously, because of the added strength and stiffness of wood structural panels on all exterior surfaces that are not window or door openings, including gable ends, narrower bracing panels may be used to give additional architectural latitude. While these reductions are permitted in IRC Table R602.10.5 (reproduced above), Figure 5.17 and Table 5.5 have been developed to clarify the issue and eliminate the need to calculate the permissible braced wall panel lengths next to openings. Figure 5.16 and Table 5.5 contain values up to 12 feet in height, as permitted in Section R301.3.

FIGURE 5.17

BRACED WALL PANELS FOR CONTINUOUS WOOD STRUCTURAL PANEL SHEATHING

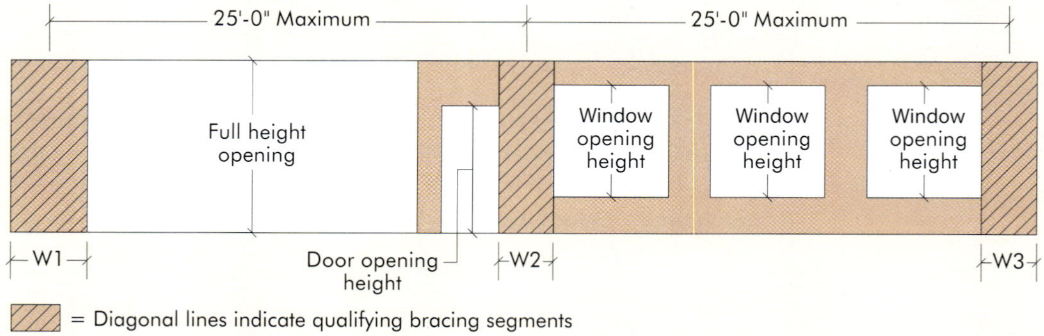

TABLE 5.5

Length of braced wall qualifying segments for continuous sheathing[a]

Wall Height (ft)	Full Height Opening 100% of Wall Height W1 (in.)	Door Opening Max. 85% of Wall Height W2 (in.)	Window Clear Opening Max. 65% of Wall Height W3 (in.)	Garage Door Opening When Supporting Roof Only
8	48	32	24	24
9	54	36	27	27
10	60	40	30	30
11[b]	66	44	33	33
12[b]	72	48	36	36

a When a window clear opening is greater than 65% of the wall height, or a door opening is greater than 85% of the wall height, interpolation to the larger size is permitted.

b. Heights of 11–12 ft for continuous sheathing are not covered in the 2006 IRC. However, IRC Section R602.10.5 requires continuous to be "in accordance with Method 3" which is permitted to be 12-ft tall. Use of continuous 12-ft-tall walls requires wider segments (for example: 12-ft-tall x 72-in. length for continuous vs. 12-ft-tall x 48 in. for isolated Method 3), because continuous sheathing wall bracing segments are aspect ratio based. Because the bracing segments are wider, and the wall is continuously sheathed, this interpretation is conservative in the opinion of the authors.

The continuous wood structural panel sheathing bracing method also has a unique length provision that applies to walls next to garage door openings. Table R602.10.5, Footnote b, provides an exception to the requirement for a braced wall panel next to a garage door opening to have a 4:1 aspect ratio. The 4:1 aspect ratio exception provided by Footnote b permits the 8-foot-high wall to be 24 inches in length.

> ***Table R602.10.5, footnote b:***
>
> *b. Full-height sheathed wall segments to either side of garage openings that support light frame roofs only, with roof covering dead loads of 3 psf or less shall be permitted to have a 4:1 aspect ratio.* (See Figure 5.18)

FIGURE 5.18

4:1 ASPECT RATIO WALL METHOD 3 – BRACING PANEL ADJACENT TO GARAGE DOOR OPENING IN CONTINUOUS WOOD STRUCTURAL PANEL SHEATHED STRUCTURE

TABLE R602.10.5, footnote b

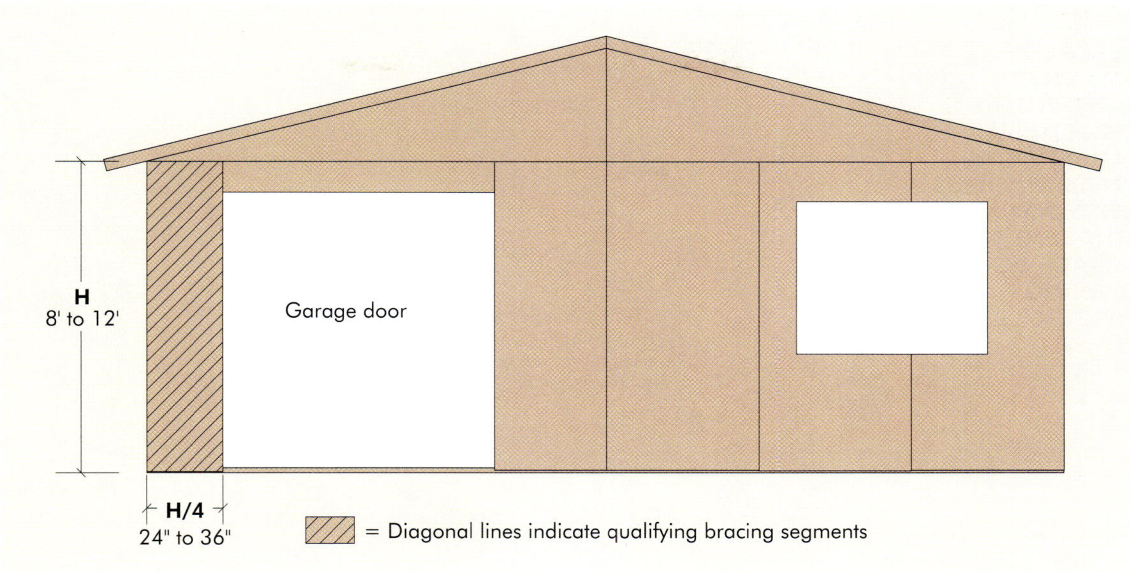

Another application that applies only to continuously sheathed wood structural panel sheathing is the allowance for 6:1 aspect ratio bracing segments next to garage doors for single story or the first of two stories in Seismic Design Categories A–C.

> ***Table R602.10.5, footnote c:***
> *c. Walls on either or both sides of openings in garages attached to fully sheathed dwellings shall be permitted to be built in accordance with Section R602.10.6.2 and Figure R602.10.6.2 except that a single bottom plate shall be permitted and two anchor bolts shall be placed at 1/3 points. In addition, tie-down devices shall not be required and the vertical wall segment shall have a maximum 6:1 height-to-width ratio (with height being measured from top of header to the bottom of the sill plate). This option shall be permitted for the first story of two-story applications in Seismic Design Categories A through C.* (See Figures 5.19 and 5.20, and Table 5.4)

FIGURE 5.19

6:1 ASPECT RATIO PORTAL FRAME WITHOUT HOLD DOWNS ADJACENT TO GARAGE DOOR OPENING IN CONTINUOUS WOOD STRUCTURAL PANEL SHEATHED STRUCTURE

TABLE R602.10.5, footnote c

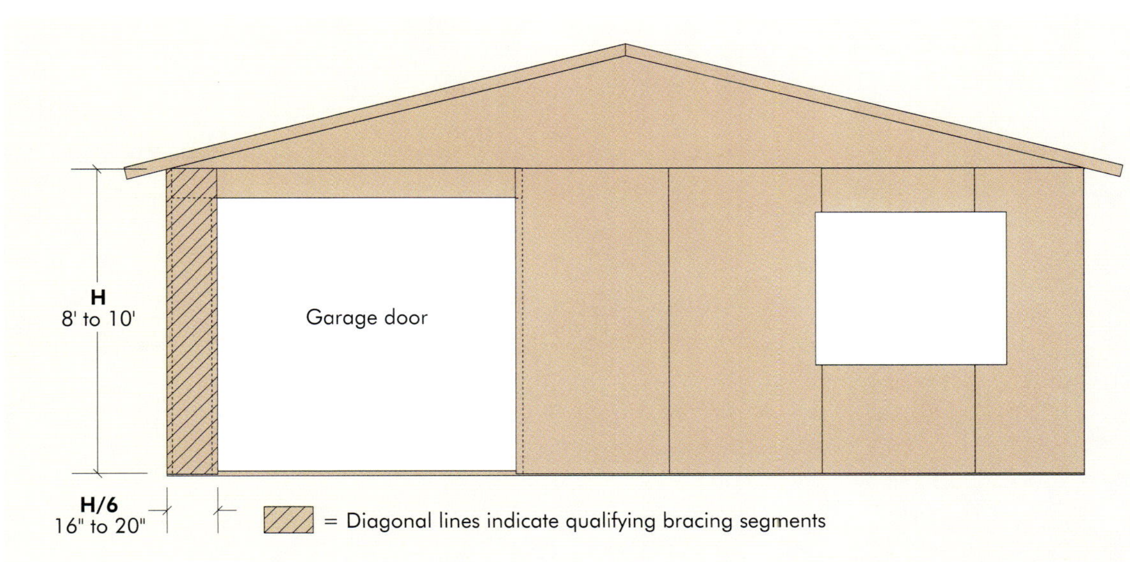

FIGURE 5.20

DETAILS OF 6:1 ASPECT RATIO WOOD STRUCTURAL PANEL PORTAL FRAME WITHOUT HOLD DOWNS ADJACENT TO A GARAGE DOOR OPENING

TABLE R602.10.5, footnote c

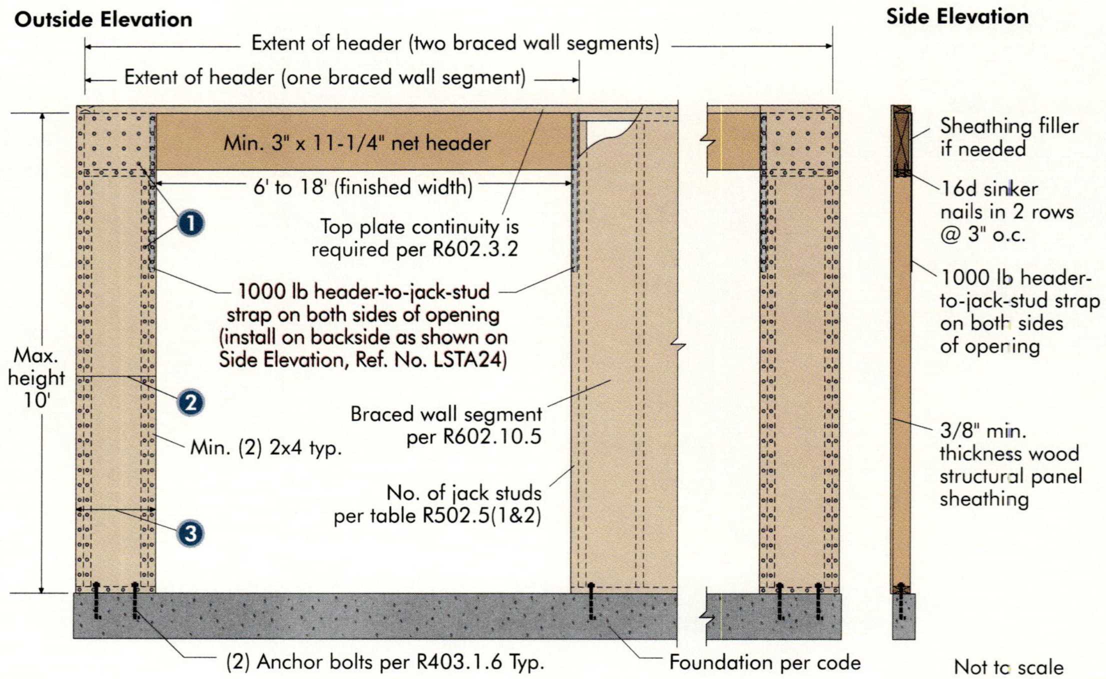

1. Fasten sheathing to header with 8d common nails in 3-in. grid pattern as shown and 3 in. o.c. in all framing (studs and sills) typ.
2. For a panel splice (if needed), panel edges shall be blocked and occur within 24 in. of mid height. One row of typ. sheathing-to-framing nailing is required in each panel.
3. Min. length based on 6:1 height-to-width ratio. Example: 16 in. min. for 8-ft height, 20-in. min. for 10-ft height, etc.

TABLE 5.6

6:1 aspect ratio portal frame widths for common wall heights

Portal Frame Length (inches)	Wall Height (ft)
16	8
18	9
20	10

Mixing bracing methods

The 2006 IRC is silent on the subject of mixing bracing methods. Mixing bracing is really three separate issues:

1. Story-to-story

Is it possible to mix bracing methods from story-to-story? An example would be using a Method 3 (wood structural panel sheathing or siding) on the bottom of a three-story building and use let-in or Method 5 (gypsum wallboard) bracing on the top story and/or second story.

2. Braced wall line to braced wall line

May a builder use different bracing methods on different walls within a story? For example, can a designer use a Method 3 along one wall line of a story to gain the benefit of a lower percent of bracing to accommodate additional windows in a "window wall," where the opening size desired on the remaining walls permits the use of Method 5 (gypsum wallboard) sheathing or some other bracing method?

3. Mixing in one braced wall line

How about the use of different bracing methods within a single braced wall line? Can a builder use let-in bracing, gypsum wallboard, and plywood on the same wall line?

Responses to these three issues

Because these issues are not covered by the IRC, the IRC (in Sections R104.11, R301.1.3, and R602.10.10) permits the use of engineering practice or testing to provide solutions to the above issues. As always in such cases, it is the local building official who has the final say on whether the evidence submitted is sufficient for acceptance.

While these issues are not covered in the IRC, as mentioned above, they are the subject of discussion within an ICC Ad Hoc Committee on Wall Bracing. At the time of this writing, the following is the consensus:

1. Variation in bracing method from story-to-story is permitted.

2. Variation in bracing method from braced wall line to braced wall line within a story is permitted, except that continuous wood structural panel sheathing shall conform to additional requirements (noted in third response on next page).

3. In Seismic Design Categories A and B, and detached dwellings in Seismic Design Category C, variation in bracing method within a braced wall line is permitted. The required sheathing percentage for the braced wall line with the mixed sheathing types shall have the higher bracing percentage, per Table R602.10.1, of all types of bracing used. Wall lines using continuous wood structural panel sheathing shall conform to the following:

a. Different bracing methods shall not be permitted within a continuously sheathed braced wall line.

b. Other bracing methods shall be permitted on other braced wall lines on the same story level or on different story levels of the building.

c. Where a continuously sheathed braced wall line is used in Seismic Design Category D_0, D_1 or D_2, or regions where the basic wind speed exceeds 100 mph, all other exterior braced wall lines on the same story shall be continuously sheathed.

TABLE 5.7

Mixing bracing methods*

Condition	Can bracing methods be mixed?
Different bracing methods, different floors	Yes
Different bracing methods, same floor	Yes
Different bracing methods, same wall	Yes, for Seismic Design Categories A–C; no, for others

* While the alternate braced wall panels in Section R602.10.6 may be mixed with other braced wall types, the alternate braced wall panels themselves must always be contructed with wood structural panels.

Other bracing methods

In addition to the eight bracing methods, two alternates and the continuous sheathed method given in the IRC, there are a large number of additional bracing options available that are recognized as code-compliant through current ICC Evaluation Service (ICC-ES) reports.

Braces and bracing methods approved for use based on an ICC-ES report must be installed in strict accordance with all details and limitations found in the ICC-ES report.

Code reports

Copies of ICC-ES and Legacy reports are available at www.icc-es.org. This site has a search function that allows you to search by report number, manufacturer's name, or product. The files are in PDF format and may be downloaded for free.

The following list is compiled to assist plan approval and construction with the bracing provisions of Chapter 6:

A. Panel width – Is each bracing panel the right minimum length?
- Method 1 (let-in bracing) – 45 to 60 degrees.
- Method 5 (gypsum board) – 8-ft gypsum on one side, 4-ft gypsum on both sides.
- All other methods – 4 ft/panel.

B. Is the wall height less than or equal to 12 ft?

C. Panel attachment – are bracing panels attached on all edges in accordance with Table R602.10.3(1), (2)?

D. Are gypsum panels, used for bracing, attached at top and bottom plates and at wall corner framing?

E. Panel grade and thickness – are bracing panels the proper grade and thickness (in accordance with R602.10.3)?

F. Is bracing percentage of wall line in accordance with Table R602.10.1?
- (Total length of braced wall panels meeting A, B, C and D above)/(total length of wall) x 100% must be greater than the % of wall line requiring bracing as required in Table R602.10.1.

G. Are code-listed bracing alternates used (R602.10.6)?
- If so, are hold downs sized and installed properly?

H. Are continuously sheathed provisions used (R602.10.5)? Is building fully sheathed with wood structural panels above and below openings, including gable ends?
- If so, is nailing used in corners in accordance with Figure R602.10.5?
- Portal frame at garage openings need not have hold downs in Seismic Design Categories A–C.
- Is portal frame built in accordance with Table R602.10.5, Footnote c?

I. Are braced wall lines spaced properly?
- Seismic Design Categories A–C less than or equal to 35 ft.
- Seismic Design Categories D_0–D_1 less than or equal to 25 ft with a single 35-ft spacing permitted adjacent to a less than or equal to 900 sf room.

J. Are interior bracing panels over a continuous foundation?
- Seismic Design Categories A–C, D_0, D_1: No foundation requirement for Seismic Design Categories A–C or for wind.
- Seismic Design Category D_2: (R602.10.9).
  1. **Single story –** max. 50-ft spacing permitted between interior continuous foundations.

(list continued on next page)

2. In two story – Are all interior braced wall lines over continuous foundations?
Continuous footing not required under all braced wall lines if:
(Any one of the following)
a. Cripple wall height less than or equal to 4 ft,
b. First floor interior braced wall lines over double floor joists, blocking, or floor beams,
c. Distance between wall lines less than or equal to 2 x building width, parallel to interior braced wall line.

K. Are interior braced wall panels properly attached top and bottom?

- All wind and Seismic Design Categories A–C – Attached in accordance with Table R602.3(1).

1. Top connection:

- If perpendicular to floor or roof framing, blocking shall be applied in line both above and below the braced wall panel and shall be attached as follows:
  1. At the top of the braced wall panel, attach with (3) 8d toenails at each block.
- If parallel to floor or roof framing but not in line with floor or roof framing, a rim joist or other parallel framing member shall be provided to facilitate attachment as follows:
  1. At the top of the braced wall panel, attach with (3) 8d toenails at each block.

2. Bottom connection: *"Sole plate to joist or blocking at braced wall panels"* – (3) 16d nails at 16 in. o.c.

- Seismic Design Categories D_0, D_1 and D_2 in accordance with Table R602.3(1) and R602.11.2.
  1. Top connection: Floor joists parallel toe-nailed with 8d nails at 6 in. o.c. to top plate. Other configurations, see "All wind and Seismic Design Categories A–C" above.
  2. Top lap splices shall have (8) 16d nails on each side of lap splice.

L. Are exterior braced wall lines properly attached? (In accordance with the requirements of Table R602.3(1).)

M. Are braced wall panels displaced from corners of building?

- Seismic Design Categories A–C – 12.5-ft displacement from corner permitted. (R602.10.1) Greater required?
  Provisions of Section R602.10.10 permit use of designed collector.
- Seismic Design Categories D_0, D_1 and D_2 – 8-ft displacement (R602.10.11.2) permitted, provided **both** of the following are satisfied:
  1. Minimum 24-in. bracing panel applied to both sides of corner at intersection walls and attached in accordance with Figure R602.10.5.
  2. 1800-lb hold down used at the end of each braced wall panel closest to the corner.

N. Are bracing panels offset from braced wall line?

- A maximum total out-to-out offset between segments making up a braced wall line of 8 ft is permitted with each offset braced wall line within a maximum of 4 ft from the designated braced wall line.

O. For dwellings in Seismic Design Categories D_0–D_2 and townhouses in any Seismic Design Category, are there irregularities per R301.2.2.2.2?

CHAPTER 6

How Much Bracing is Needed?

The amount of wall bracing required in each braced wall line, as specified by the building code, depends on the Seismic Design Category (SDC) or wind speed, the stories above the braced wall line, and the method of bracing used.

Braced segments, called braced wall panels, must occur near each end of the braced wall line and at least every 25 feet o.c. They must also make up a certain percentage of the braced wall line. Table 6.1 summarizes the bracing requirements.

TABLE 6.1

Bracing requirements (For complete details see Table R602.10.1 of the IRC)

Seismic Design Category (SDC) or Wind Speed	Stories Above Braced Wall Line[a]	Method of Bracing Permitted[b,c]	Amount of Bracing Per Wall Line: For Method 3 Bracing	Amount of Bracing Per Wall Line: For Other Methods Permitted	Maximum Spacing Between Braced Wall Lines (ft)
SDC A and B or winds 100 mph and less	0	Methods 1–8	16%	16%	35[d]
	1	Methods 1–8	16%	25%	
	2	Methods 2–8	25%	35%	
SDC C or winds less than 110 mph	0	Methods 1–8	16%	25%	
	1	Methods 2–8	30%	45%	
	2	Methods 2–8	45%	60%	
SDC D_0, D_1 or winds less than 110 mph	0	Methods 2–8	20%	30%	25[e]
	1	Methods 2–8	45%	60%	
	2	Methods 2–8	60%	85%	
SDC D_2 or winds less than 110 mph	0	Methods 2–8	25%	40%	
	1	Methods 2–8	55%	75%	

a. Stories above braced wall line.

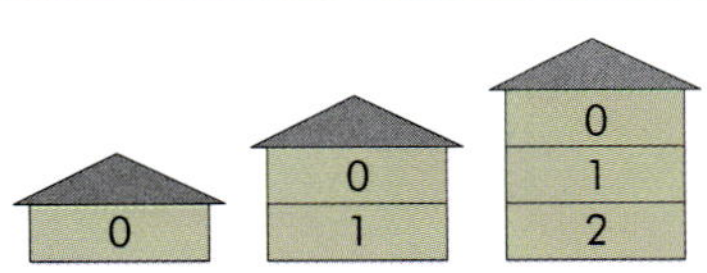

0 = one story or top of two- or three-story house
1 = first story of two-story or second story of three-story house
2 = first story of three-story house

b. Continuous structural panel sheathing (IRC R602.10.5) is permitted everywhere.

c. Alternate braced wall segments (IRC R602.10.6) are permitted in one story and the first story of two-story houses, and require continuous foundations.

d. See Table 6.2 of this book and IRC Section R602.10.1.1 for exceptions.

e. See Chapter 3 of this book and IRC Section R602.10.11 for exceptions.

Adjustments to the required amount of bracing

As with many code requirements, there are adjustments and exceptions that apply to the requirements. The adjustments in Table 6.2 have been collected from various sections of the IRC and put together in a single table below.

TABLE 6.2

Adjustment factors to the amount of required wall bracing[a]

ADJUSTMENT BASED ON:			MULTIPLY AMOUNT OF BRACING PER WALL LINE BY:	APPLIES TO:
Story height[b] (Section R301.3)		10 ft or less	1.0	All bracing methods
		12 ft	1.2	
Braced wall line spacing in SDC A-C[b,d] (Section R602.10.1.1)		35 ft or less	1.0	
		50 ft	1.43	
Wall dead load[e] (Section R301.2.2.2.1)		8 – 15 psf	1.0	
		Less than 8 psf	0.85	
Roof/ceiling dead load for wall supporting[b,c]: (Table R301.2.2.2.1)	roof only or roof plus one story	15 psf or less	1.0	
	roof only	25 psf	1.1	
	roof plus one story	25 psf	1.2	
Continuous wood structural panel sheathing (R602.10.5)		Opening height 0.67 x h or less	0.8	Continuous wood structural panel sheathing
		Opening height 0.85 x h or less	0.9	
		Opening height equal to h	1.0	
Walls with stone or masonry veneer in SDC C–D_2		See Section R703.7, Exceptions 1–2		
Cripple walls		See Section R602.10.2		

a. The total amount of bracing required for a given wall line is the product of all applicable adjustment factors.
b. Linear interpolation shall be permitted.
c. Bracing required for a site's wind speed shall not be adjusted for dead load.
d. Braced wall line spacing in excess of 35 ft shall be in accordance with R602.10.1.1.
e. The adjusted amount of bracing shall not be less than that required for the site's wind speed.

Clarification of opening height adjustment for continuous wood structural panel bracing

The code allows for reductions in the percentage of bracing required in a wall line when using continuous wood structural panel sheathing (IRC R602.10.5). When using continuous wood structural panel sheathing as bracing, the amounts of bracing required, as described in Table 6.1, may be adjusted by a combination of factors given in Table 6.2. For continuous wood structural panel sheathing, the bracing amount may also be adjusted by a factor of:

- 0.9 for walls with openings of 0.85 multiplied by the wall height and less, or
- 0.8 for walls with openings of 0.67 multiplied by wall height and less

This is shown in Figure 6.1. These factors are only applicable to continuous wood structural panel sheathing and not to other bracing methods.

FIGURE 6.1

When using continuous wood structural panel sheathing, the required amount of sheathing may be further reduced.

Maximum opening size in wall line

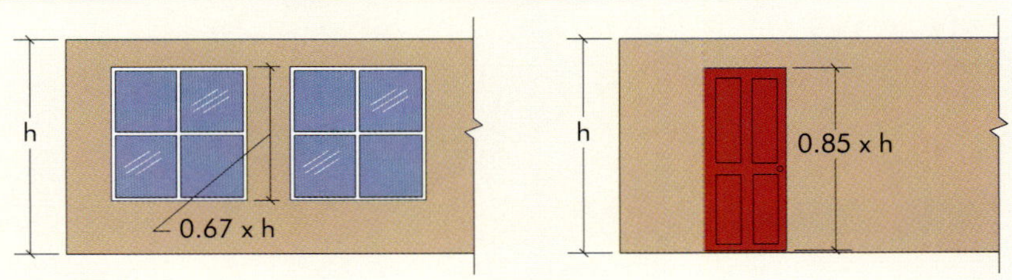

Amount of bracing reduction factor for wall line	0.8	0.9
Example	If the wall line in the figure above requires 25% bracing and the wall is braced per R602.10.5 and the maximum window clear height in the wall line is no larger than 0.67 x h, then the 25% shall be permitted to be reduced to 25% x 0.8 = 20%	If the wall line in the figure above requires 45% bracing and the wall is braced per R602.10.5 and the maximum opening clear height in the wall line is no larger than 0.85 x h, then the 25% shall be permitted to be reduced to 45% x 0.9 = 40.5%

How is the percentage of bracing in a wall line determined?

The bracing percent in a given wall line is simply the sum of the length of the qualifying braced wall segments, divided by the total length of the braced wall line and multiplied by 100 percent. For example, the percentage of bracing in braced wall line C in Figure 6.2 is the sum of the two bracing segments E and F, divided by the length L and multiplied by 100 percent.

FIGURE 6.2

DETERMINING THE AMOUNT OF BRACING, EXPRESSED AS A PERCENT, IN A BRACED WALL LINE

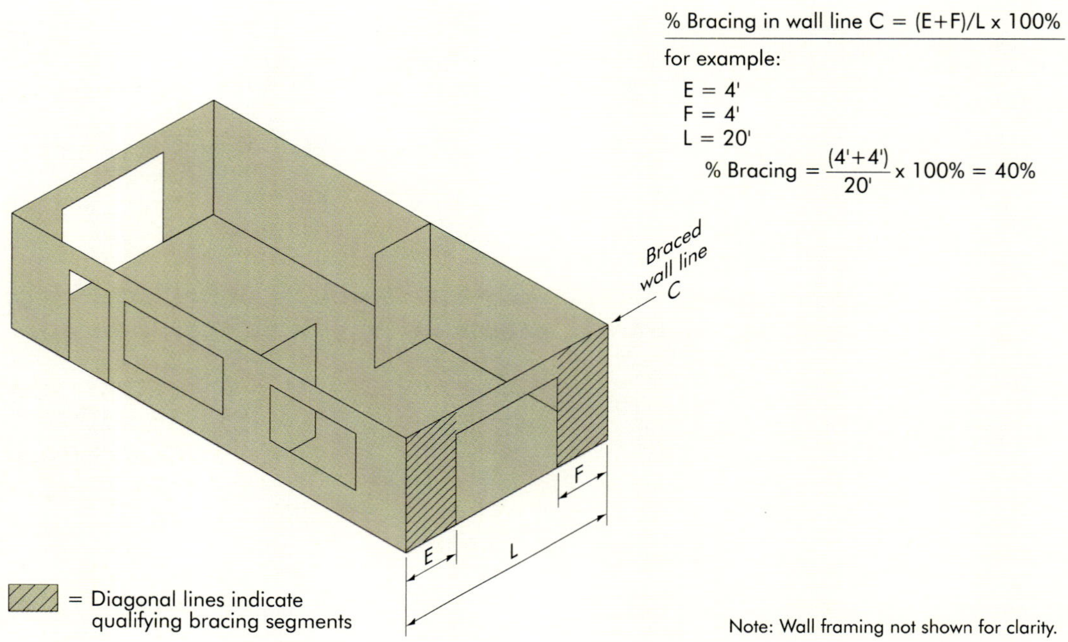

Gypsum board when used as bracing (Method 5)

Section R602.10.4 of the 2006 IRC, requires each gypsum board bracing panel to be 96 inches in length when applied to one face of a braced wall line. When applied to both faces, the minimum required length becomes 48 inches.

What the 2006 IRC doesn't provide is guidance on how to compute percent bracing for these two situations. This issue was considered by the ICC Ad Hoc Committee on Wall Bracing. It was determined by this committee that both 96 inches of single-sided Method 5 bracing and 48 inches of double-sided Method 5 bracing would be considered 48 inches for computing percent bracing required by Table R602.10.1. The Ad Hoc Committee supported a code change proposal during the 2006–07 code-change cycle to make this clear. This proposal was approved during the final Action Hearing in May of 2007.

Computing percent bracing examples

Example 6.1: Example for computing percent bracing using structural fiberboard (Method 4) (See Figure 6.3)

How much bracing is required and present in a given wall line using Method 4 bracing?

Given:

- The house is in Seismic Design Category A, with 105 mph winds.
- 48-inch structural fiberboard (Method 4) bracing is used.
- Braced wall line has no story above it.
- Figure 6.3.

Solution:

Step 1. Determine how much bracing is required for given scenario.

- From Table 6.1 (or IRC Table R602.10.1) we can see that bracing is required at each end of the braced wall line and at least every 25 feet, but not less than 25 percent due to 105 mph wind speed.

Step 2. Determine how much qualified bracing is present in the braced wall line.

- $\frac{(48 \text{ inches} + 48 \text{ inches})}{12 \text{ inches per foot}} = 8 \text{ feet}$

- $\frac{8 \text{ ft}}{24 \text{ ft}} \text{ x } 100\% = 33.3\%$

Step 3. Since 33.3 percent of bracing is greater than the 25 percent required, this braced wall line meets the required amount of bracing for the example.

FIGURE 6.3
EXAMPLE USING STRUCTURAL FIBERBOARD (METHOD 4)

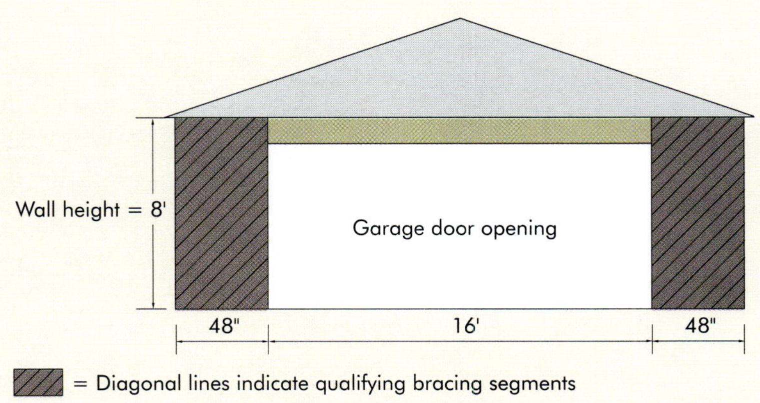

Example 6.2: Example for computing percent bracing using particleboard (Method 6) in 10-foot-tall wall (See Figure 6.4)

How much bracing is required and present in a given wall line (Method 6) bracing?

Given:

- The house is in Seismic Design Category A, with 90 mph winds.
- 48-inch particleboard (Method 6) bracing is used.
- Braced wall line has one story above it.
- Figure 6.4.

Solution:

Step 1. Determine how much bracing is required for given scenario.

- From Table 6.1 (or IRC Table R602.10.1) we can see that bracing is required at each end of the braced wall line and at least every 25 feet, but not less than 25 percent.

Step 2. Determine how much qualified bracing is present in the braced wall line.

- $\frac{(48 \text{ inches} + 48 \text{ inches})}{12 \text{ inches per foot}} = 8 \text{ feet}$
- $\frac{8 \text{ ft}}{24 \text{ ft}} \text{ x } 100\% = 33.3\%$

Step 3. Since 33.3 percent of bracing is greater than the 25 percent required, this braced wall line meets the required amount. Notice that the 24- and 32-inch wall sections were not counted as bracing because they do not meet the 48-inch minimum width requirement for this bracing method.

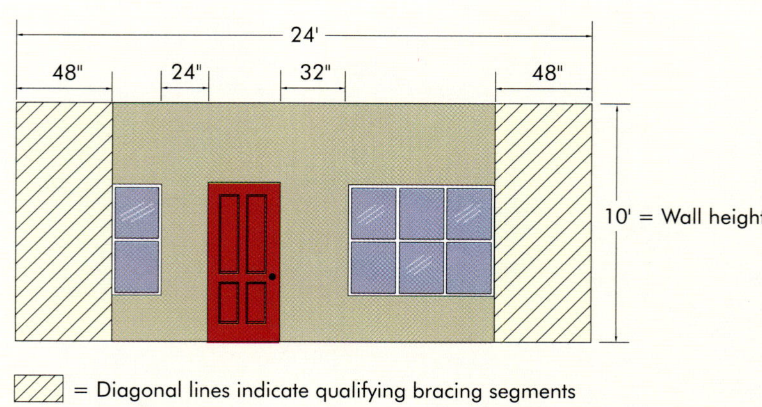

FIGURE 6.4

EXAMPLE USING PARTICLEBOARD (METHOD 6) IN 10-FOOT-TALL WALL

Example 6.3: Example for computing percent bracing using hardboard panel siding (Method 8), away from corner, in 9-foot-tall wall with offsets in braced wall line (See Figure 6.5)

Given:

- The house is in Seismic Design Category A, with 90 mph winds.
- 48-inch-wide hardboard siding (Method 8) bracing is used.
- Braced wall line has one story above it.
- Figure 6.5.

Solution:

Step 1. Determine how much bracing is required for given scenario.

- From Table 6.1 (or IRC Table R602.10.1) we can see that bracing is required at each end of the braced wall line and at least every 25 feet, but not less than 25 percent.

Step 2. Determine how much qualified bracing is present in the braced wall line.

- $\dfrac{(48 \text{ inches} + 48 \text{ inches} + 48 \text{ inches})}{12 \text{ inches per foot}} = 12 \text{ feet}$
- $\dfrac{12 \text{ ft}}{40 \text{ ft}} \times 100\% = 30\%$

Step 3. Since 30 percent of bracing is greater than the 25 percent required, this braced wall line meets the required amount of bracing for the given conditions.

FIGURE 6.5

EXAMPLE USING HARDBOARD PANEL SIDING (METHOD 8), AWAY FROM CORNER, IN 9-FOOT-TALL WALL WITH OFFSETS IN BRACED WALL LINE

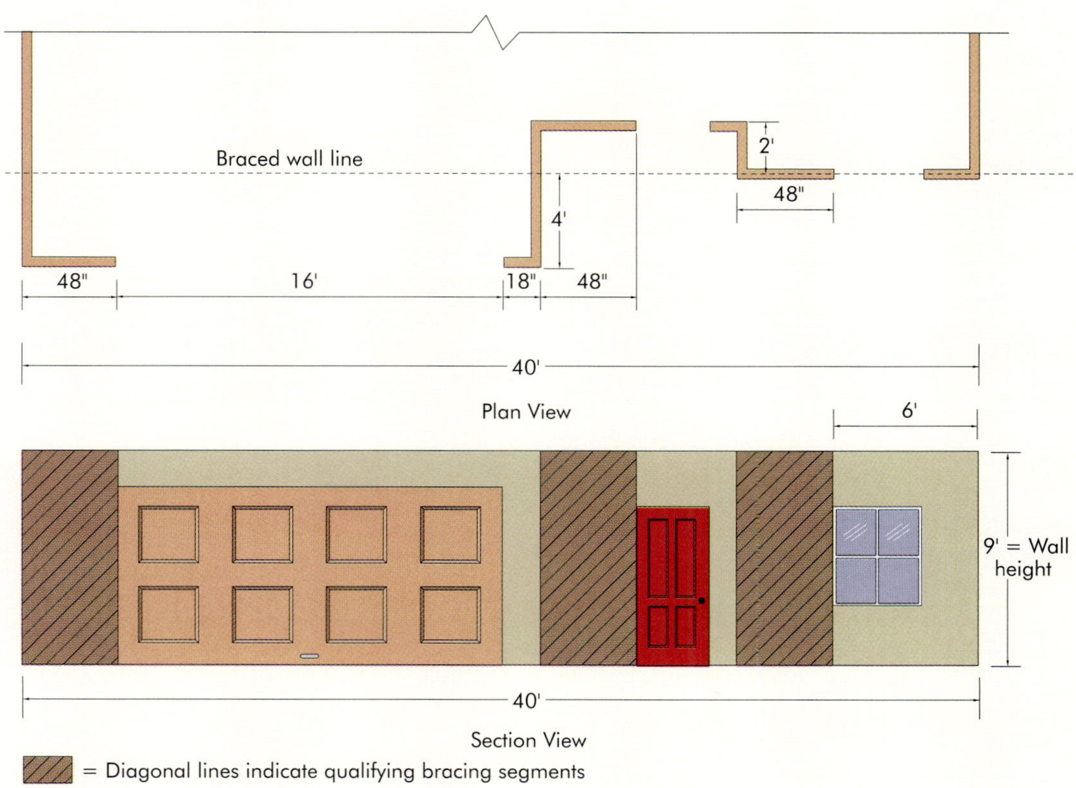

Example 6.4: Example for computing percent bracing using portland cement plaster (Method 7) bracing

(See Figure 6.6)

Given:

- The house is in Seismic Design Category A, with 105 mph winds.
- Portland cement plaster (Method 7) bracing is used.
- Braced wall line has one story above it.
- Figure 6.6.

Solution:

Step 1. Determine how much bracing is required for given scenario.

- From Table 6.1 (or IRC Table R602.10.1) we can see that bracing is required at each end of the braced wall line and at least every 25 feet, but not less than 45 percent.

Step 2. Determine how much qualified bracing is present in the braced wall line.

- (7 ft + 6 ft + 7 ft) = 20 feet
- $\frac{20 \text{ ft}}{36 \text{ ft}} \text{ x } 100\% = 55.6\%$

Step 3. Since 55.6 percent of bracing is greater than the 45 percent required, this braced wall meets the required amount of bracing for the given conditions.

FIGURE 6.6
EXAMPLE USING PORTLAND CEMENT PLASTER (METHOD 7) BRACING

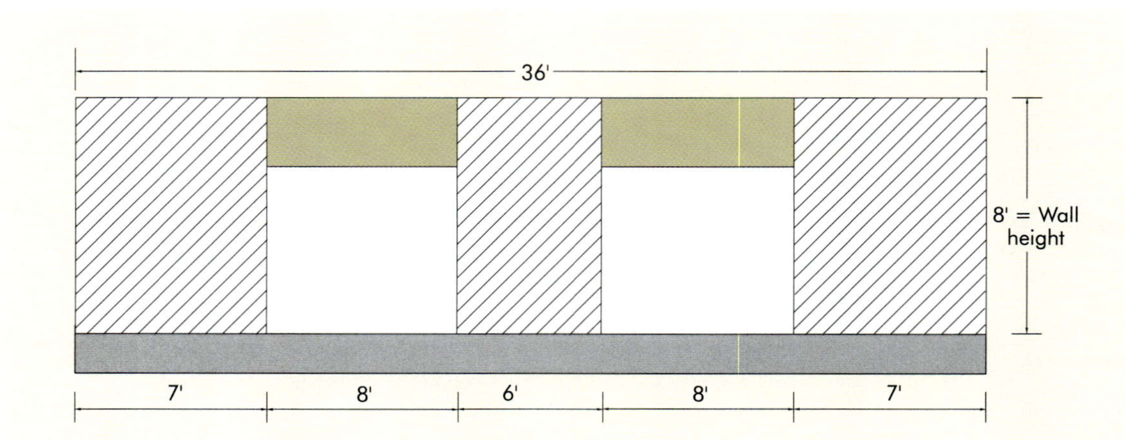

Computing percentage bracing when using narrow wall (less than 48 in.) braced wall panels

Example 6.5: Garage exception (See Figure 6.7)

Footnote b of Table R602.10.5 permits bracing segments on either side of single story garages to be 4:1 height-to-length ratio. This bracing option is part of R602.10.5 and can only be used when walls are continuously sheathed with wood structural panels in accordance with R602.10.5.

Given:

- The garage is located in Seismic Design Category C, with 100 mph winds.
- Continuous wood structural panel sheathing is used per R602.10.5.
- The braced wall line has a light frame roof with dead loads 3 psf or less on the roof.
- Figure 6.7.

Solution:

Step 1. Determine how much bracing is required for the given scenario.

- From Table 6.1 (or IRC Table R602.10.1) we can see that bracing is required at each end of the braced wall line and at least every 25 feet, but not less than 16 percent.

Step 2. Determine how much qualified bracing is present in the given braced wall line in Figure 6.3.

- $\frac{(2\text{ ft} + 2\text{ ft})}{20\text{ ft}} \times 100\% = 20\%$

Step 3. Since 20 percent of bracing is greater than 16 percent, this braced wall line meets the required amount for a one-story continuously sheathed garage.

FIGURE 6.7
4:1 ASPECT RATIO WALL SEGMENTS AT GARAGE DOORS

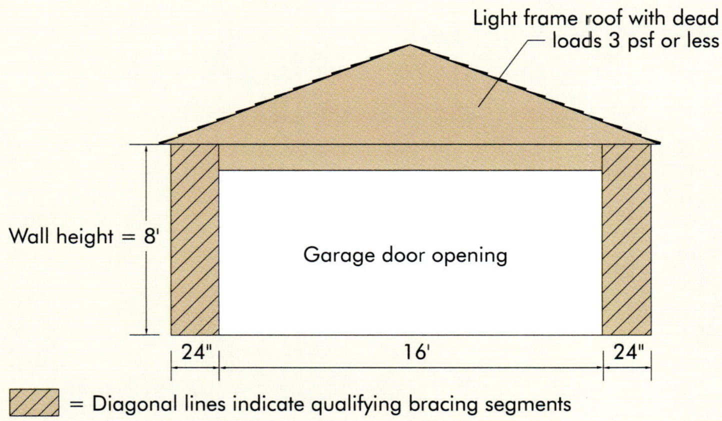

Example 6.6: 24-inch-wide bracing segments in a braced wall line

(See Figure 6.8)

Given:

- The house is in Seismic Design Category B, with 100 mph design wind speeds.
- Continuous wood structural panel sheathing is used per R602.10.5.
  (See also Table R602.10.5.)
- The braced wall line has one story above it. (Not shown in the figure.)
- Figure 6.8.

Solution:

Step 1. Determine how much bracing is required for given scenario.

- From Table 6.1 (or IRC Table R602.10.1), we can see that bracing is required at each end of the braced wall line and at least every 25 feet, but not less than 16 percent.

- The 16 percent can be reduced to 16% x 0.9 = 14.4% or 16% x 0.8 = 12.8% for continuously sheathed walls with openings limited as stated.

- For the braced wall line in this example, the minimum percentage of bracing is governed by the door height, thus the minimum amount of bracing is 16% x 0.9 = 14.4%. Note that the 24-inch braced wall segment adjacent to the door is too narrow to count.

Step 2. Determine how much qualified bracing is present in the braced wall line.

- $\dfrac{(24 \text{ inches} + 32 \text{ inches} + 24 \text{ inches})}{12 \text{ inches per foot}} = 6.67 \text{ feet}$

- $\dfrac{6.67 \text{ ft}}{24 \text{ ft}} \times 100\% = 27.7\%$

Step 3. Since 27.7 percent of bracing is greater than the 14.4 percent required, this braced wall line meets the required amount.

FIGURE 6.8

24-INCH-WIDE WOOD STRUCTURAL PANEL BRACING SEGMENTS IN A BRACED WALL LINE

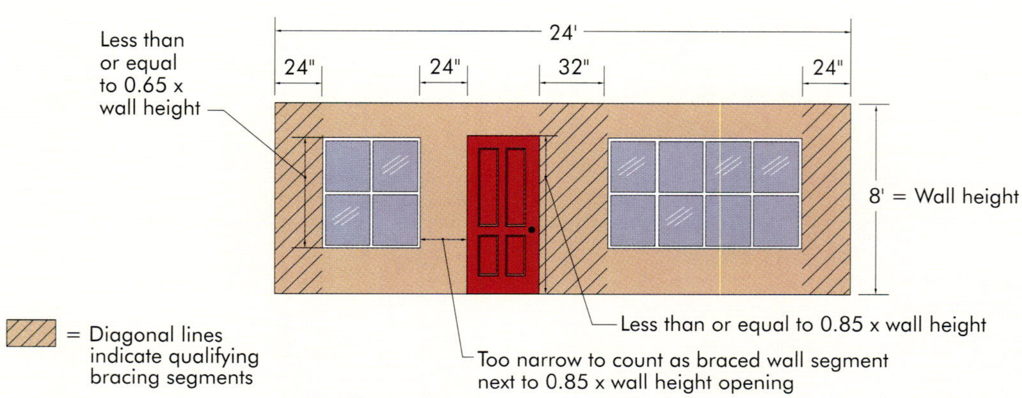

Example 6.7: Narrow wall bracing at a garage braced wall line (See Figure 6.9)

Given:

- The house is in Seismic Design Category B, with 100 mph design wind speeds.
- Continuous wood structural panel sheathing is used per R602.10.5 and Table R602.10.5, Footnote c.
- Braced wall line has one story above it. (Not shown in the figure.)
- Note that Footnote c of Table R602.10.5 permits the use of 6:1 aspect ratio segments when adjacent to garage door openings. For a 10-foot-tall wall, the minimum length of this segment is 20 inches as discussed in Chapter 5.
- Figure 6.9.

Solution:

Step 1. Determine how much bracing is required for given scenario.

- From Table 6.1 (or IRC Table R602.10.1), we can see that bracing is required at each end of the braced wall line and at least every 25 feet, but not less than 16 percent.

Step 2.
- 6:1 Aspect ratio
- 10 feet = 120 inches
- $\frac{120 \text{ inches}}{6} = 20\text{-inch wall length}$

Step 3. Determine how much qualified bracing is present in the braced wall line.

- $\frac{(20 \text{ inches} + 20 \text{ inches} + 20 \text{ inches})}{12 \text{ inches per foot}} = 5 \text{ feet}$
- $\frac{5 \text{ ft}}{29 \text{ ft}} \text{ x } 100\% = 17.2\%$

Step 4. Since 17.2 percent of bracing is greater than the 16 percent required, this braced wall line meets the required amount for the ground floor of a two-story building.

FIGURE 6.9
NARROW WALL WOOD STRUCTURAL PANEL BRACING AT A GARAGE BRACED WALL LINE

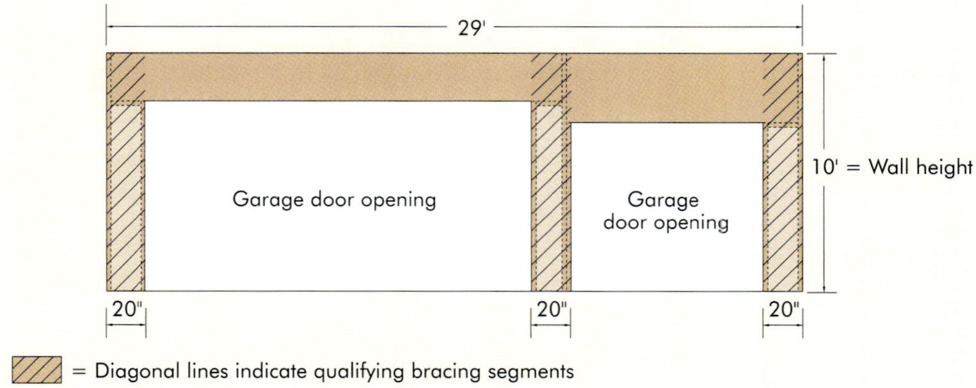

Example 6.8: Attached garage with offsets in braced wall line (See Figure 6.10)

Given:

- The house is in Seismic Design Category A, with 90 mph design wind speeds.
- Continuous wood structural panel sheathing is used per R602.10.5 and Table R602.10.5, Footnote c.
- Braced wall line has one story above it. (Not shown in figure.)
- Note that Footnote c of Table R602.10.5 permits the use of 6:1 aspect ratio segments when adjacent to garage door openings. For a 9-foot-tall wall, the minimum length of this segment is 18 inches as discussed in Chapter 5.

  Table R602.10.5 permits the use of a 27-inch-wide segment in a 9-foot-tall wall adjacent to a wall opening that is less than or equal to 0.65 x the wall height. In this case, 0.65 x 9 feet is a wall opening height of 5.85 feet. The given window height is 5 feet, which is less than 5.85 feet, therefore, 27 inches is appropriate.
- Figure 6.10.

Solution:

Step 1. Determine how much bracing is required for the garage in the given scenario.

- From Table 6.1 (or IRC Table R602.10.1), we can see that bracing is required at each end of the braced wall line and at least every 25 feet, but not less than 16 percent.

Step 2.

- 6:1 Aspect ratio
- 9 feet = 108 inches
- $\frac{108 \text{ inches}}{6} = 18\text{-inch wall length}$

Step 3. Determine how much qualified bracing is present in the braced wall line.

- $\frac{(18 \text{ inches} + 18 \text{ inches} + 27 \text{ inches} + 27 \text{ inches})}{12 \text{ inches per foot}} = 7.5 \text{ feet}$
- $\frac{7.5 \text{ ft}}{40 \text{ ft}} \times 100\% = 18.8\%$

Step 4. Since 18.8 percent of bracing is greater than the 16 percent required, this braced wall line meets the required amount for the ground floor of a two-story building (note that if the maximum opening height were 0.85 multiplied by the wall height, or less, the 16 percent required could be reduced by a 0.9 factor to 14.4 percent).

FIGURE 6.10

ATTACHED GARAGE WITH OFFSETS IN CONTINUOUS WOOD STRUCTURAL PANEL BRACED WALL LINE

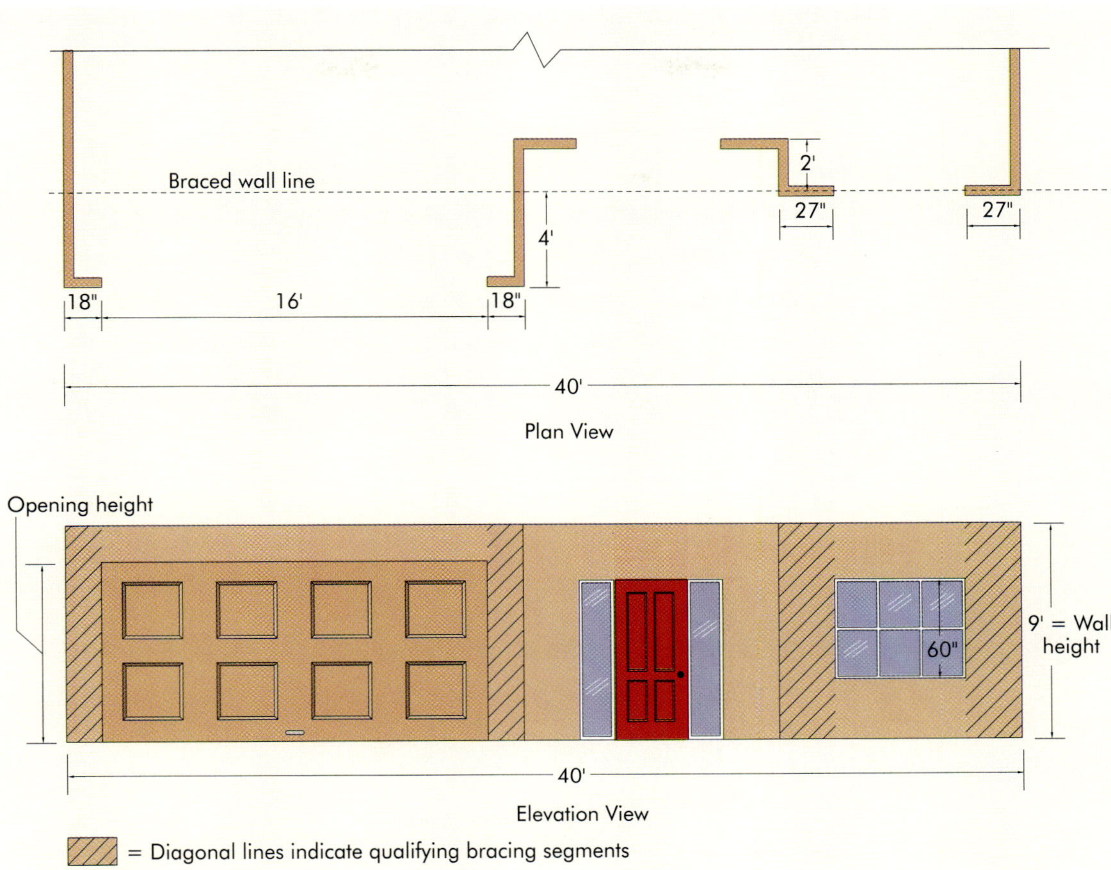

Example 6.9: Example using 32-inch alternate with hold downs (See Figure 6.11)

How much bracing is required and present in a given wall line using 32-inch-wide alternate bracing with hold downs?

Given:

- The house is in Seismic Design Category B, with 105 mph winds.
- 32-inch alternate bracing with hold downs with wood structural panel sheathing on both faces is used per R602.10.6.1.
- Braced wall line has one story above it. (Not shown in figure.)
- Figure 6.11.

Solution:

Step 1. Determine how much bracing is required for given scenario.

- From Table 6.1 (or IRC Table R602.10.1), we can see that bracing is required at each end of the braced wall line and at least every 25 feet, but not less than 30 percent.

Step 2. Determine how much qualified bracing is present in the braced wall line.

- Note that each 32-inch segment, for the purpose of computing the percentage of bracing, is equivalent to 48 inches of bracing.

$$\frac{(48 \text{ inches} + 48 \text{ inches})}{12 \text{ inches per foot}} = 8 \text{ feet}$$

- $\frac{8 \text{ ft}}{21.33 \text{ ft}} \times 100\% = 37.5\%$

Step 3. Since 37.5 percent of bracing is greater than the 30 percent required, this braced wall line meets the required amount of bracing for the given conditions.

FIGURE 6.11
EXAMPLE USING 32-INCH WOOD STRUCTURAL PANEL ALTERNATE WITH HOLD DOWNS

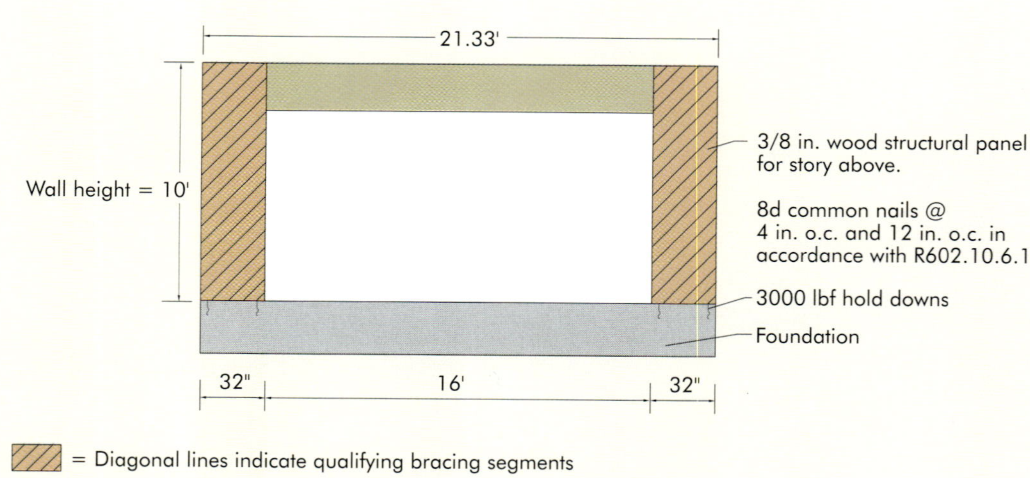

Example 6.10: Example using 24-inch alternate portal frame with hold downs

(See Figure 6.12)

How much bracing is required and present in a given wall line using 24-inch-wide alternate portal frame bracing with hold downs?

Given:

- The house is in Seismic Design Category D_2, with 105 mph winds.
- 24-inch alternate bracing with hold downs is used per R602.10.6.2.
- Braced wall line has one story above it. (Not shown in figure.)
- Figure 6.12.

Solution:

Step 1. Determine how much bracing is required for given scenario.

- From Table 6.1 (or IRC Table R602.10.1), we can see that bracing is required at each end of the braced wall line and at least every 25 feet, but not less than 55 percent.

Step 2. Determine how much qualified bracing is present in the braced wall line.

- Note that each 24-inch segment, for the purpose of computing the percentage of bracing, is equivalent to 48 inches of bracing.
- $\frac{(48 \text{ inches} + 48 \text{ inches} + 48 \text{ inches})}{12 \text{ inches per foot}} = 12 \text{ feet}$
- $\frac{12 \text{ ft}}{24 \text{ ft}} \times 100\% = 50\%$

Step 3. Since 50 percent of bracing is not greater than the 55 percent required, this braced wall line does not meet the required amount. This example illustrates how difficult it is to meet bracing requirements in high seismic locations. For this wall line, additional bracing length would have to be provided or an engineered solution used.

FIGURE 6.12

EXAMPLE USING 24-INCH WOOD STRUCTURAL PANEL ALTERNATE PORTAL FRAME WITH HOLD DOWNS

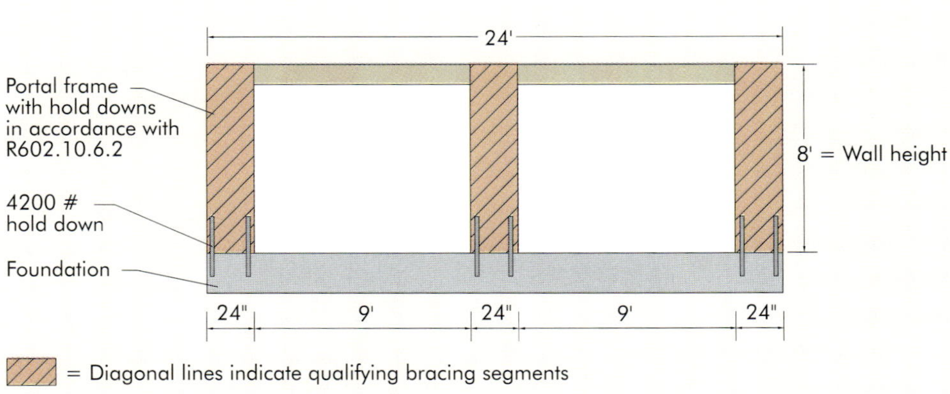

CHAPTER 7

Bracing Away from Corners, Collectors and Wall Line Offsets

This chapter describes common bracing complexities, such as bracing away from corners, understanding collectors, and determining offsets in braced wall lines.

Bracing segments away from the end of braced wall lines

Bracing segments may occur away from the end of a braced wall line and still be code compliant. For all bracing methods in Seismic Design Categories (SDC) A–C, except continuous sheathing (IRC R602.10.5), bracing panels may begin no more than 12.5 feet from the end of the wall line, as shown in Figure 7.1. If bracing is located more than 12.5 feet from the end, an engineered collector is needed to transfer lateral loads to the bracing segments in the braced wall line. More detailed information on design of a collector will be presented in this chapter.

FIGURE 7.1
BRACED WALL SEGMENTS CAN BEGIN UP TO 12.5 FEET FROM THE END OF A BRACED WALL LINE IN SDC A–C

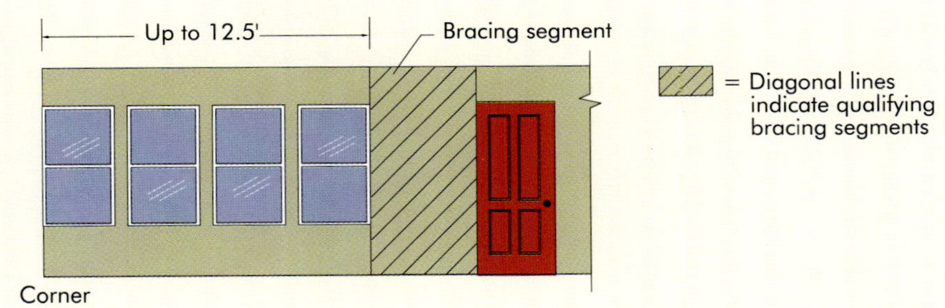

For continuous wood structural panel bracing, a corner, as shown in Figure R602.10.5, is required at the ends of braced wall lines. The 2006 IRC doesn't specify what minimum length of wood structural panel sheathing is required at the corners, but it is the position of the IRC AD HOC Committee on Wall Bracing that 2 feet should be the minimum required. It is likely that future editions of the IRC will clarify this length.

Bracing segments away from the end of braced wall lines in Seismic Design Categories D_0–D_2

For all structures in Seismic Design Categories D_0–D_2, per IRC R602.10.11, a braced wall segment must occur at the end of a braced wall line for all bracing methods except Method 3 (wood structural panel bracing). Wood structural panel bracing may be placed up to 8 feet from the end, provided one of the following provisions is met, as shown in Figure 7.2:

- A minimum 1800 lbf tie-down device is required on the end of each braced wall segment closest to the corner, or
- A minimum 24-inch-wide segment is at the corners.

FIGURE 7.2

FOR SDC D_0–D_2, TWO OPTIONS EXIST FOR BRACING AWAY FROM CORNERS

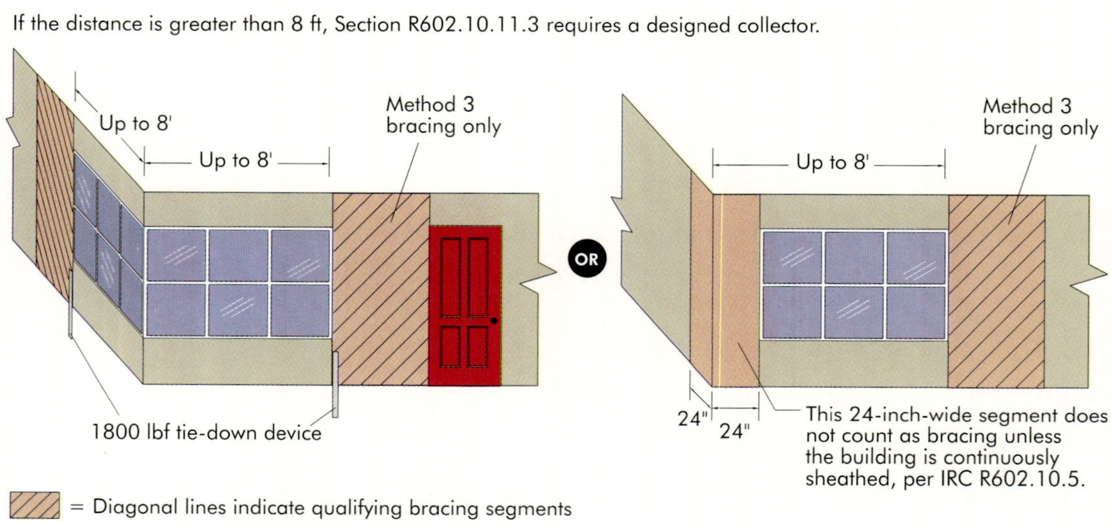

Collectors

When bracing is placed farther from the end of the braced wall line than specified (8 feet for high seismic bracing (Method 3 only), or 12.5 feet for all bracing methods elsewhere), the limits of the IRC are exceeded and, thus, require the engineered design of a collector (see R602.10.10. and R602.10.11.3).

What is a collector and what does it do?

The collector referenced in the IRC is a part of the lateral load path.

Just as a beam in the vertical load path accumulates (or "collects") gravity loads and carries them over an opening to supporting members on each side, the collector has a similar function in the lateral load path. It collects the lateral loads from the roof or floor diaphragm (roof sheathing and floor sheathing, respectively) above and distributes them into the braced wall panels on either side of this opening in the braced wall line below. Thus, a collector is needed to evenly distribute the load among the bracing panels in a given wall line that may have window and door openings that do not allow the minimum bracing requirement to be met.

What does a collector look like and where do I get one?

In the type of structures considered by the IRC, the collector is normally already in place in the form of the wall double top plate or top plate/header combination. See Figure 7.4.

In terms of load path, the roof or upper-floor framing is attached to the wall top plate. The top plate is also a part of the framing for the bracing panels immediately below. This connection between the roof or floor framing and the wall ensures that the horizontal lateral loads (from the floor or roof sheathing/diaphragm above) are distributed into the bracing panels (vertical lateral load resisting elements). The IRC Table R602.3(1) attachment requirement for double top plate splices (Minimum 24-inch offset with eight 16d nails in splice area) ensures continuity over the braced wall line up to the 8- or 12.5-foot corner distance limit. Beyond these limits, the prescriptive top plate splice cannot be counted on to transfer greater loads. The IRC-required fastener schedule, per Table R602.3(1), for the wall bottom plate of the bracing panel (three 16d face nails into joist or blocking every 16 inches o.c.) completes the load path to the foundation or floor below.

Collector design

The good historical performance of traditional framing methods in conventional construction has led to provisions in the codes that permit the first bracing panel to be placed away from the corner up to a set distance without requiring the builder to be concerned with the collector design. For distances in excess of those permitted by the code, the material properties of a single 2x4 top plate are generally adequate to distribute the applied load for conventional construction (approximate capacity for a single SPF 2x4 "Stud Grade" in tension is 3,230 lbf, based on the relatively short duration of lateral loads, such as wind and earthquake). The difficulty lies in providing sufficient attachment between the upper and lower top plates (or the upper top plate and header below) to transfer the load at "splice locations" where joints in the double top plate occur. (See Figure 7.3.)

Note that the IRC requires that the splice nailing, as illustrated in Figure 7.3, be placed on both sides of the top plate joint. (Section R602.11.2)

FIGURE 7.3
SPLICE AT TOP PLATE REQUIRED TO TRANSFER LOAD ACROSS JOINT IN UPPER TOP PLATE TO LOWER PLATE AND BACK UP INTO THE UPPER TOP PLATE AFTER THE SPLICE

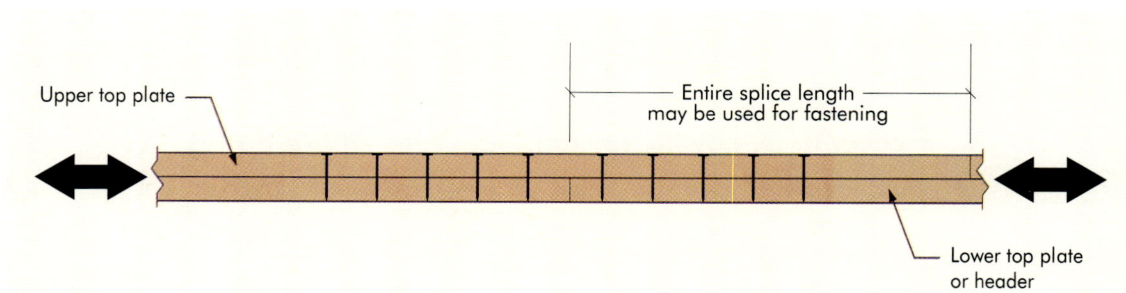

What is the length of the top plate that must be spliced?

Since the purpose of the collector is to "collect" the load from the structure above and transfer it into the bracing panel away from the corner, the spliced top plate must run from the building corner to the far side of the bracing panel. For example, if the edge of the corner bracing panel nearest the corner is to be relocated 14 feet from the corner, and the bracing panel is 4 feet wide, the total length of the spliced top plate must be a minimum of 18 feet (14 feet + 4 feet).

The required number of nails for the top plate splice can be calculated as follows:

$$N = (W \times L)/Z'$$

Where: N = Number of nails required at "splice"
W = Maximum load on the diaphragm (lbf/ft)
L = Distance between corner and first bracing panel (ft)
Z′ = Allowable capacity of a single fastener (lbf/fastener)

W: The maximum load that can be expected to be transferred into the upper top plate. This load is the *greater* of the roof diaphragm capacity or the rim-board-to-top-plate connection capacity:

- Roof diaphragm capacity: Based on 3/8-inch Rated Sheathing (RS), 8d nails at 6- and 12-inch unblocked diaphragm (Case 1) – **215 plf.**
- Rim-board-to-top-plate: Based on ICC-ES AC 124 for conventional construction – **180 plf.**

Use W = **215 plf.**

Z′: (Based on a 16d box nail (0.135 x 3-1/2 inches) joining 2 SPF top plates)
Z = 88 lb/nail (2005 National Design Specification Table 11N, and
1.6 is the duration of load adjustment for short term, wind or seismic loads.
Z′ = 88 x 1.6 = **141 lbf/nail.**

Given the above, the splice fastener requirement reduces to:

$N = (215/141) \times L$

$N = 1.52L$ **Use 1.5L (16d box nails) on each side of the butt joint in the upper top plate, where L is the distance between the corner and first bracing panel in feet.**

The top plate splice design parameters can be represented in tabular form, as shown in Table 7.1.

TABLE 7.1

Top plate splice design table[a,b]

<table>
<tr><td colspan="2">L (Distance between corner and beginning of first bracing panel in ft)</td><td>8[d]</td><td>10[d]</td><td>12[d]</td><td>12.5[d]</td><td>14</td><td>16</td><td>18</td><td>20</td></tr>
<tr><td colspan="2">Load (Maximum load on splice for conventional construction in lbf)</td><td>1720</td><td>2150</td><td>2580</td><td>2688</td><td>3010</td><td>3440</td><td>3870</td><td>4300</td></tr>
<tr><td colspan="2">N (Number of 16d box nails required on each side of top plate butt joint (nails)</td><td>12</td><td>15</td><td>18</td><td>19</td><td>21</td><td>24</td><td>27</td><td>30</td></tr>
<tr><td rowspan="2">Minimum Size, Species, and grade[c]</td><td>2x4</td><td colspan="5">SPF Stud Grade</td><td colspan="2">DF Stud Grade or SPF #2</td><td>DF #2 or SPF #2</td></tr>
<tr><td>2x6</td><td colspan="8">SPF Stud Grade</td></tr>
</table>

a. If 16d **common** nails (0.162 x 3-1/2 in.) are used, the number of nails in the above table may be multiplied by 0.73.

b. If a single top plate is used, a metal plate tension tie may be used to make the top plate splice. It must be sized in accordance with the "LOAD" row in the table above.

c. Once a size, species, and grade is selected from the table, other sizes, species and grades listed to the right of the selected grade may be substituted.

d. Engineered collector not required under the 2006 IRC for wind or SDCs A–C when bracing begins within 12-1/2 ft of end of wall.

Design example:

A home designer wants to place a bracing panel 14 feet away from a corner. The wall framing is 2x4. Select the species, grade and design, and detail the collector. See Figure 7.4.

FIGURE 7.4

DESIGN EXAMPLE SHOWING FIRST BRACING PANEL 14 FEET FROM CORNER

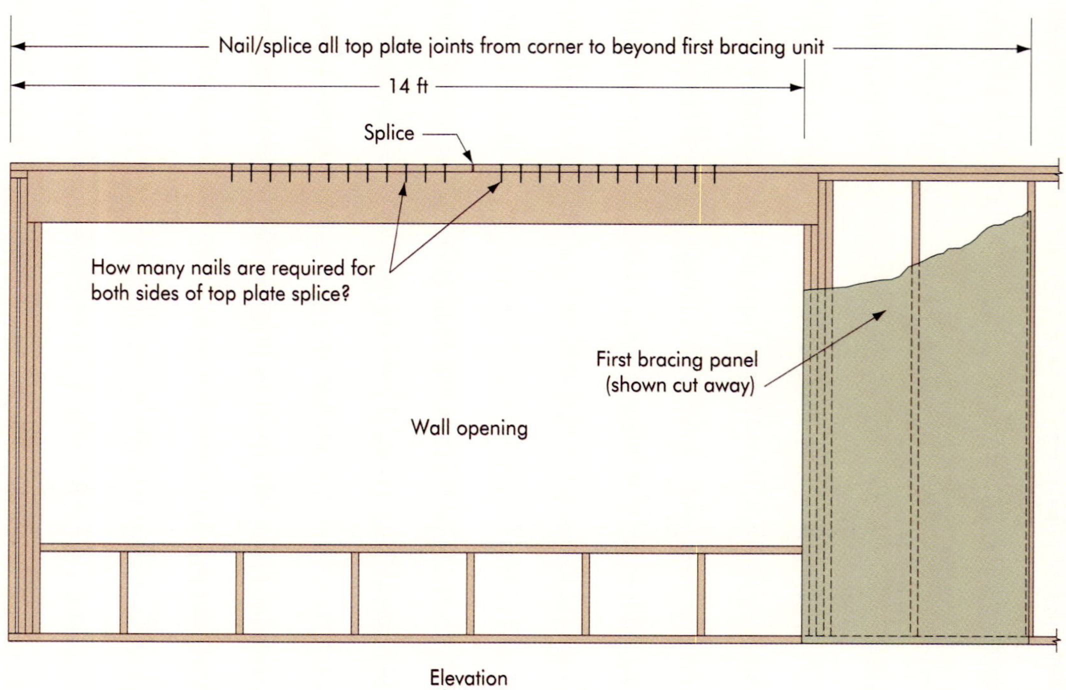

Solution: For an L of 14 feet, from Table 7.1, the top plate splice made of 2x4 spruce-pine-fir stud grade requires 21 (1.5 x L = 1.5 x 14 = 21) 16d box nails. If common nails are used, 16 (21 x 0.73 = 15.33) 16d common nails are required. Where the splice occurs over a lower top plate, a minimum 4-foot overlap would, for this example, provide enough room for required fasteners without causing splitting of the plates. Where the splice occurs over a header, a lower top plate is not necessary.

Note that using longer plates (12–16 feet) minimizes the number of plate splices required.

The splice is to be detailed as shown below in Figure 7.5.

FIGURE 7.5

TOP PLATE SPLICE DETAIL

TWENTY-ONE 16D BOX NAILS SHOWN EACH SIDE OF JOINT

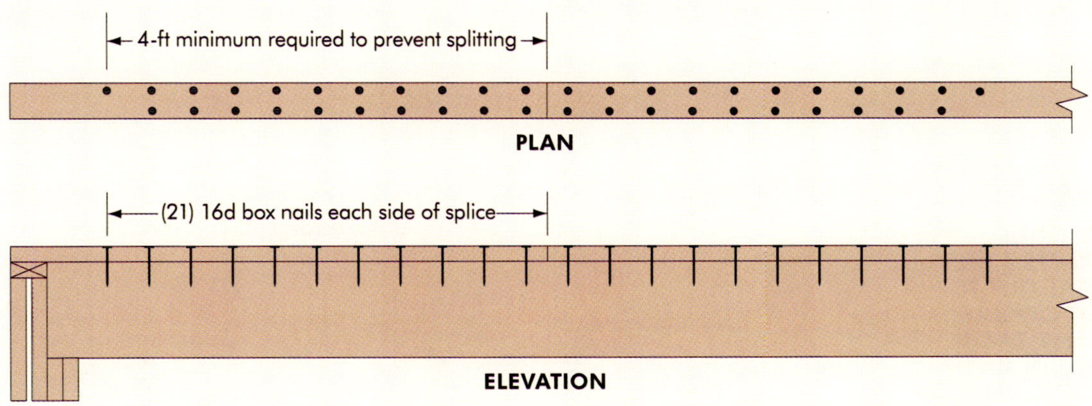

Please note that *theoretically*, 4 feet would be enough splice overlap to prevent splitting of the plates. We recommend staggering each row of nails, blunting the tips of the nails and/or using three rows of fasteners if splitting appears to be a problem given the wood resource used.

Offsets in a braced wall line

Many home designs feature offsets along the wall length. IRC R602.10.1 permits offsets up to 4 feet, provided that the total out-to-out offset dimension is not greater than 8 feet, as shown in Figures 7.6 through 7.8.

FIGURE 7.6

A BRACED WALL LINE CAN HAVE 4-FOOT OFFSETS

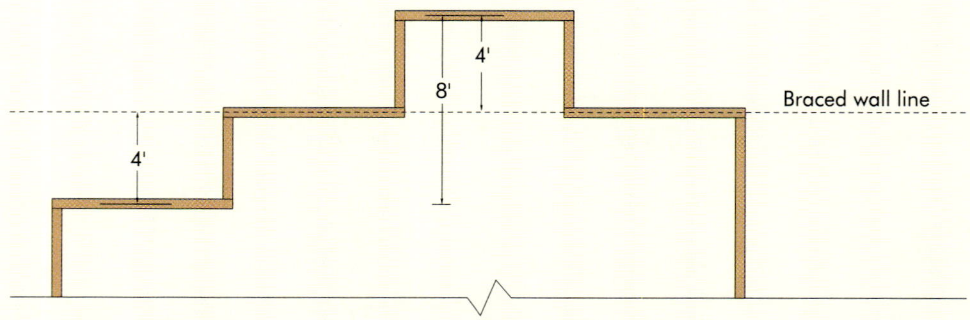

FIGURE 7.7

THE CODE PERMITS 8 FEET TOTAL OUT-TO-OUT OFFSETS (4 FEET EACH WAY) IN A BRACED WALL LINE

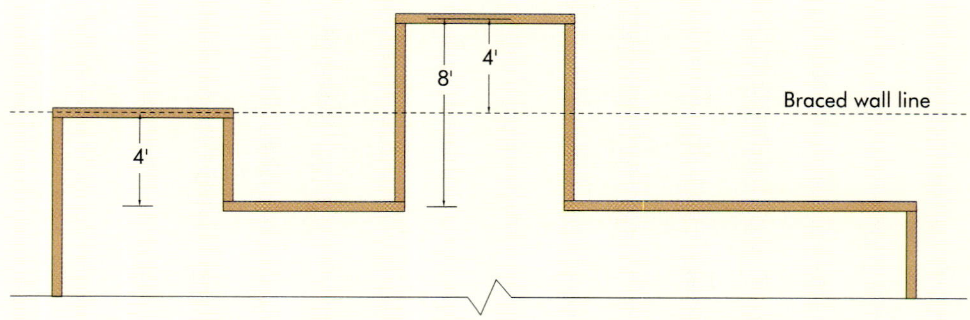

FIGURE 7.8

OFFSETS MAY ALSO OCCUR IN DISCONTINUOUS BRACED WALL LINES

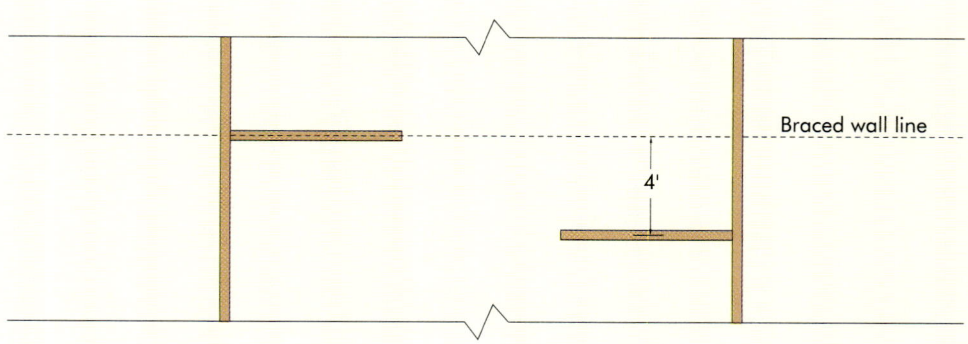

Imaginary braced wall lines (not covered in code)

While the concept of imaginary braced wall lines is sometimes used, it is not in the IRC. This concept cannot be used unless specifically approved by the building official as an alternative design under Section R104.11. The concept is based on the principle that when a number of braced wall segments run parallel to each other, all within 8 feet, it is ***not necessary*** to have an actual bracing segment on the braced wall line. The imaginary braced wall line uses the center of the 8-foot grouping, whether or not a physical wall line exists, as shown in Figure 7.9. This center is the imaginary braced wall line. The location of this imaginary wall line is used when measuring distances between the other parallel braced wall lines.

FIGURE 7.9
IMAGINARY BRACED WALL LINE

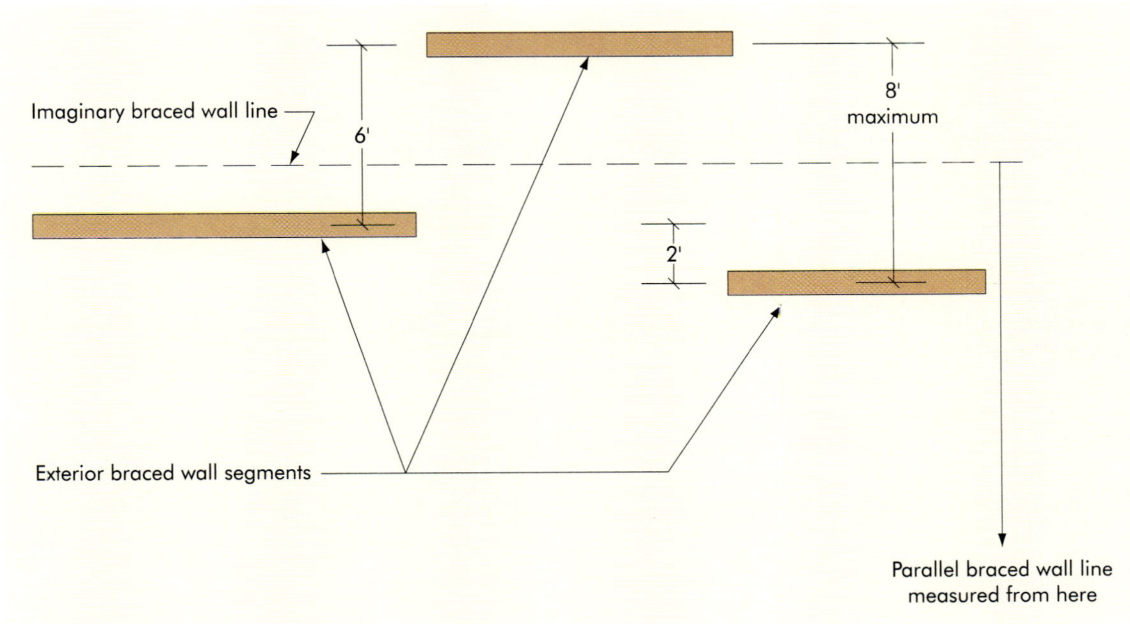

CHAPTER 8

Bracing Connection Details

This chapter describes the IRC connection requirements between the braced wall segments and the structure. Without the required connections (load path), the bracing will not function. In addition to IRC requirements, other information and suggestions for proper application and reasonable solutions will be provided.

Braced wall segments to raised wood floor connections – wind less than 110 mph and Seismic Design Categories A–C

In Table R602.3(1), the IRC requires that the sole plate (or bottom plate) of braced wall segments be nailed with three 16d nails every 16 inches o.c. into the joist or blocking below braced wall panels.

Section R602.10.8 requires that braced wall lines running parallel to the floor framing be provided with "a rim joist or other parallel framing members" to facilitate the attachment required by Table R602.3(1) and shown below.

DESCRIPTION OF BUILDING ELEMENTS	MEMBER AND TYPE OF FASTENER	SPACING OF FASTENERS
Sole plate to joist or blocking, at braced wall panels	3-16d (3-1/2" x 0.135")	16" o.c.

Where joists are perpendicular to the braced wall lines above, blocking must be provided under and in line with the braced wall segment. The braced wall segment top plate must be connected to the framing above in accordance with Figure 8.1, IRC Table R602.3(1) and R602.10.8.

FIGURE 8.1

REQUIREMENTS TO CONNECT BRACING TO THE STRUCTURE ON A RAISED WOOD FLOOR

(IRC TABLE R602.3(1) AND SECTION R602.10.8)

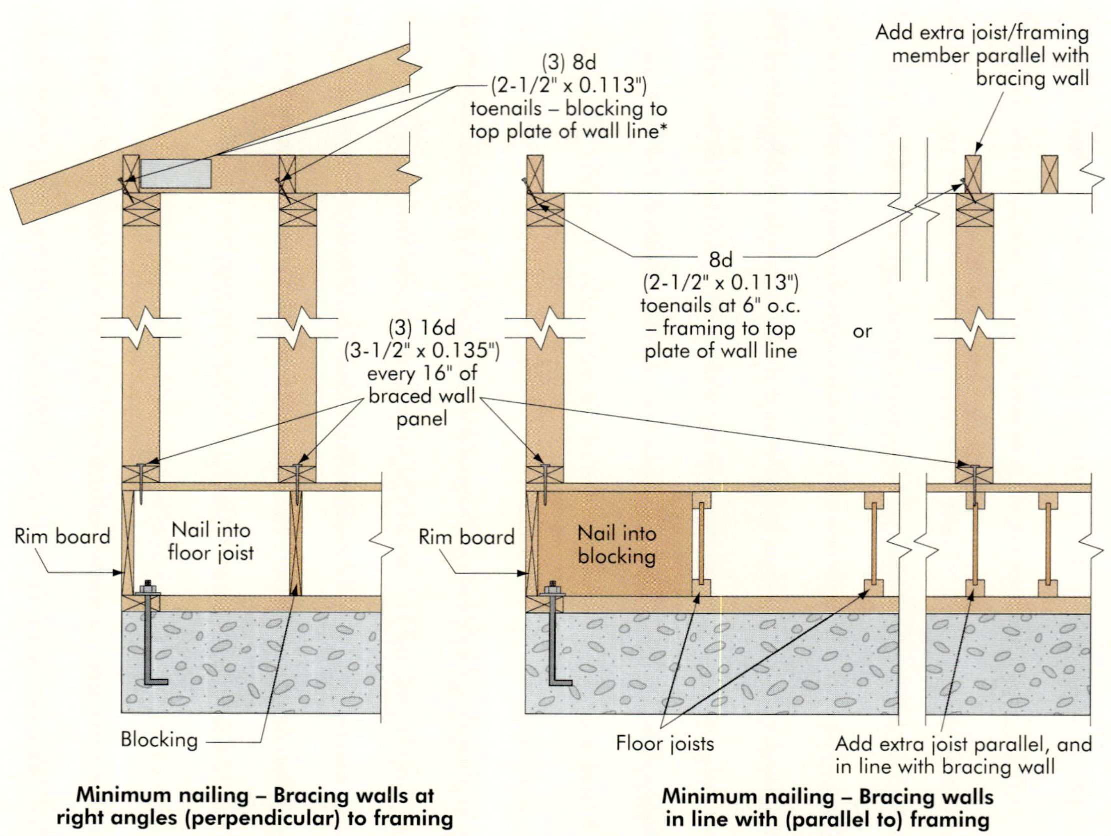

* Section R602.10.8 requires that braced wall lines running parallel to the floor framing or roof/ceiling framing be provided with "a rim joist or other parallel framing members" to facilitate the attachment required by Table R602.3(1). In Seismic Design Categories D_0 and higher, a nailng pattern specified in Section R602.11.2 is provided.

Bracing anchored directly on a foundation – wind less than 110 mph and Seismic Design Categories A–C

When braced wall panels are supported directly on foundations (e.g., slab on grade, masonry or concrete foundation wall, etc.), the wood sill plate must be anchored to the foundation with anchor bolts spaced a maximum of 6 feet o.c. There must be a minimum of two bolts per sill plate section with one bolt located not more than 12 inches or less than seven bolt diameters (3.5 inches) from each end of the plate section (IRC R403.1.6). Bolts should be at least 1/2 inch in diameter and should extend a minimum of 7 inches into the concrete or masonry foundation. A nut and washer is required on each bolt to hold the plate to the foundation. See Figure 8.2.

FIGURE 8.2

REQUIREMENTS TO CONNECT SILL PLATES TO THE FOUNDATION

(IRC R403.1.6)

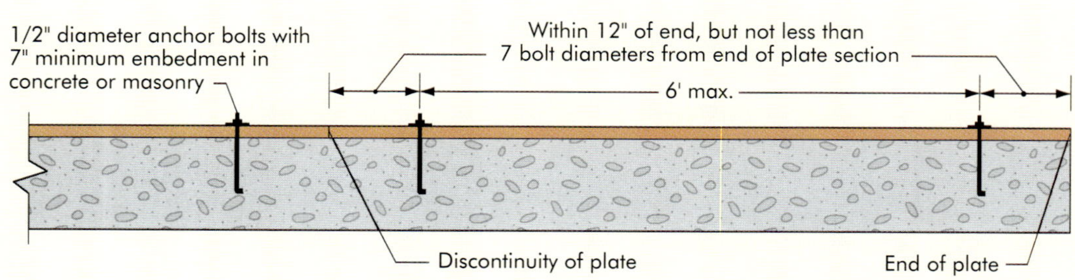

Additional requirements for high seismic

The following provisions are required in addition to all of the requirements for Seismic Design Categories (SDC) A–C described previously:

- For all buildings in Seismic Design Categories D_0, D_1 and D_2, and townhouses in Seismic Design Category C, the plate washers are required to be 3 gage (0.229 inch) x 3 x 3 inches (IRC R602.11.1).

- In addition to all of the previously discussed requirements, **interior braced wall panels** in Seismic Design Categories D_0, D_1 and D_2 require:

1. Floor joists parallel to the braced wall segment top plate must be toenailed to the top plate with at least 8d nails spaced a maximum of 6 inches o.c. (IRC R602.11.2), as seen in Figure 8.3.

FIGURE 8.3
INTERIOR BRACED WALL ATTACHMENT TO FLOOR JOISTS IN SDC D_0, D_1 AND D_2
(R602.11.2)

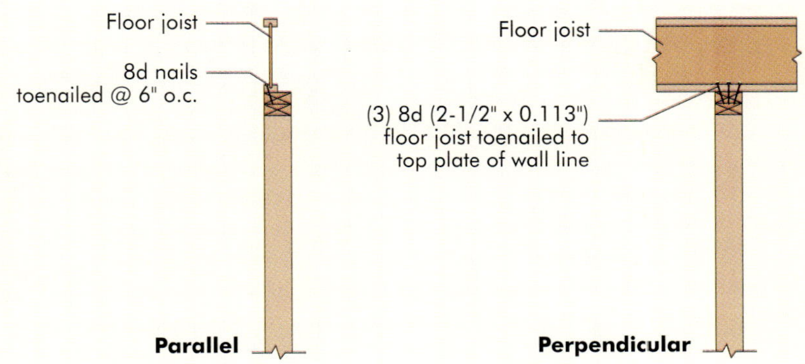

FIGURE 8.4
TOP PLATE SPLICE AS FOR SDC D_0, D_1 AND D_2
(R602.11.2)

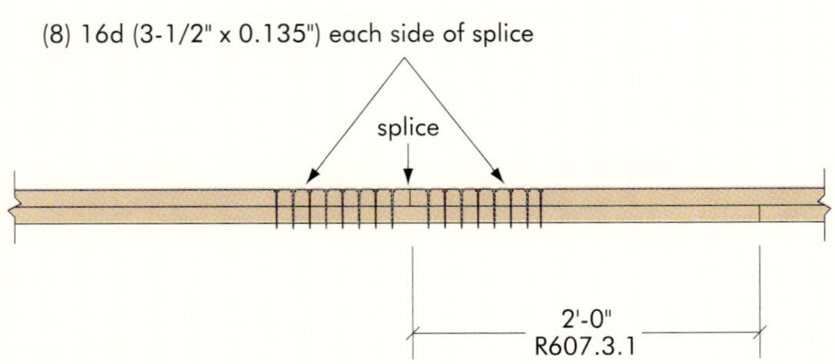

Top plate splice detail for interior braced wall line

2. Top plate lap splices must be face nailed with at least eight 16d nails on each side of the splice to maintain continuity throughout the length of the interior braced wall line, as shown in Figure 8.4.

Foundations and attachment requirements for braced wall lines

In general, the foundation is the last major component of a structure's vertical and lateral load paths.

For **vertical loads,** the foundation must be designed to distribute the vertical load (the weight of the structure and loads acting on it) into the soil below the foundation. This is discussed in Chapter 1 of this book. When properly designed, the load imparted on the soil by the structure must be less than or equal to what the soil can support. The heavier the structure, the wider the foundation has to be to distribute the load into the soil. Similarly, a poorer quality soil requires a wider footing.

The **lateral loads** on a structure are also distributed into the soil as the last step in the load path. Lateral events, such as earthquakes and high winds, generate two separate forces that must be transferred into the soil. These forces are sliding and overturning. Sliding forces, discussed in Chapter 2, try to push the structure along the surface of the ground. In structures covered by the IRC, the foundation extends below the surface of the ground and resists sliding forces by a combination of friction against the surrounding soil surface and the compressive resistance of the soil acting against the foundation in the direction of the lateral load. If the structure is adequately connected to the foundation, and the foundation extends below the surface of the ground as specified by the IRC, this is usually not a problem.

In addition to the sliding force, a second force generated by earthquake or high wind is an overturning force, also discussed in Chapter 2. As the lateral load acts on the structure above ground level, a "tipping" or overturning force is generated.

An example of this phenomenon would be trying to move a refrigerator across the floor by pushing the top. As long as the refrigerator can roll along without any resistance there is no problem. But, as soon as the refrigerator hits the carpet, and the bottom stops moving, that force at the top tries to tip the refrigerator over.

Anchoring the foundation in the soil to prevent sliding secures the foundation, thereby, resisting the tipping force or overturning. The foundation resists this force in two ways. On the side opposite of the lateral load, the footing is prevented from rotating downward by the soil's resistance, as discussed for vertical loads. On the side with the acting lateral load, the foundation tries to pull out of the ground. This upward force is resisted by the friction of the soil on each side of the foundation, the weight of the soil on top of the footing, the weight of the structure and the weight of the foundation itself.

FIGURE 8.5

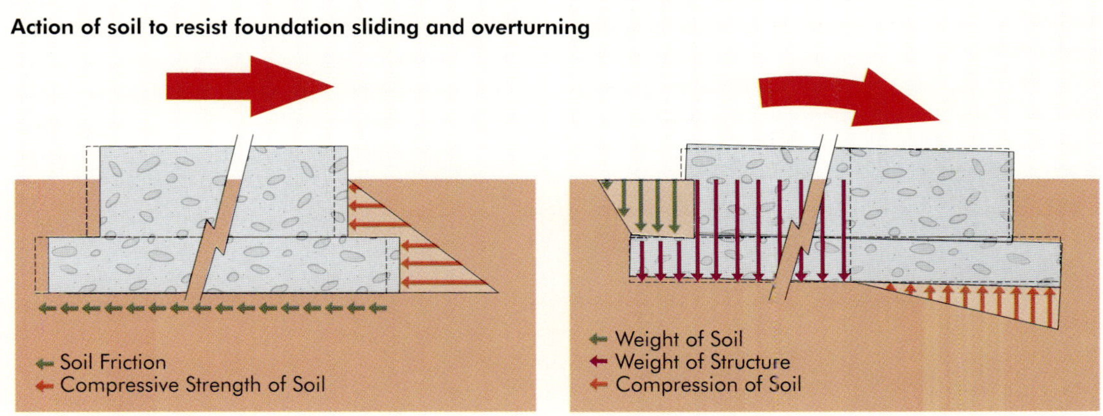

Lateral Load
Resisted by soil friction and compression capacity of soil.

Overturning
Resisted by weight of structure, weight of soil pushed up and compression capacity of soil.

If these forces are high and the structure is well attached to the foundation, the foundation must often be designed to be very deep and/or designed with a wide footing. This wide footing transfers upward forces into a wedge of soil on each side of the footing. If the footing is going to pull out, it has to lift that wedge of soil with it. This effectively increases the weight of the foundation and, thus, its overturning resistance.

Continuous footings, for both interior and exterior braced wall lines, are an important part of this discussion because they are the only type of foundation prescriptively required (Sections R403.1.2, R403.1.3, R403.1.6.1 and R602.10.9) or described (R403.1.6, R602.10.8 and R602.11.1) in the IRC. There are a number of other foundation types that may be built per R602.10.10. These include piles, piers, floating foundations and discontinuous foundations; however, because this publication concerns itself with the IRC, the content will be limited to continuous footings. In addition, this publication will focus on light frame construction to be consistent with the bracing provisions found in Chapter 6 of this document.

Continuous footings. The IRC restrictions on size and load limit the actual foundation loads in these types of structures. The vertical loads are limited by prescriptive requirements in IRC Chapter 3 (R301.2). In a similar manner, the lateral loads are accommodated by IRC Chapter 4 (R403) prescriptive requirements and through the wall anchorage requirements of IRC Chapter 6 (R602.11.1). (See Figure 8.6.) In Seismic Design Categories D_0, D_1 and D_2, additional reinforcement requirements for foundations are specified, as noted in Figure 8.7.

FIGURE 8.6
MINIMUM EXTERIOR BRACING WALL FOUNDATION REQUIREMENTS

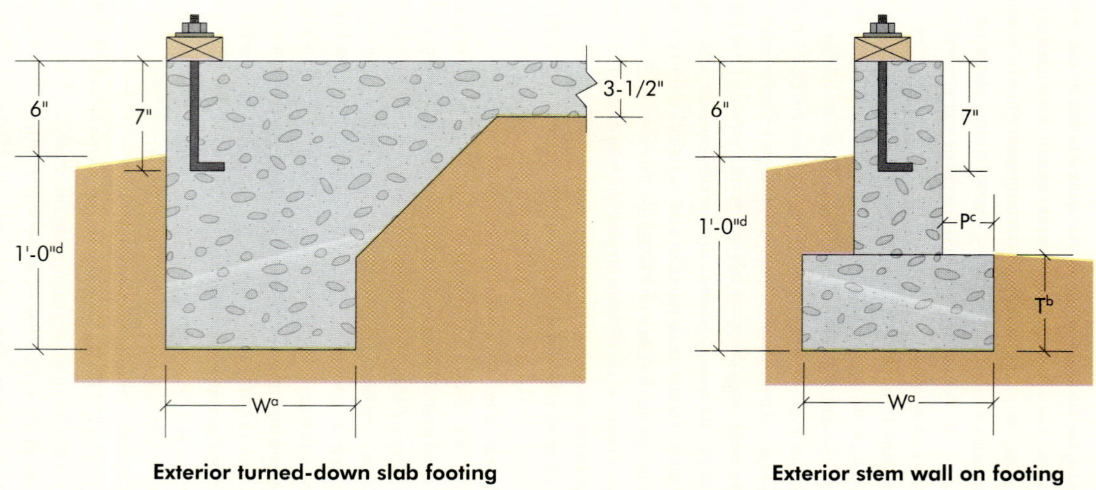

a. W = per Table R403.1.
b. T = 6 in. min. per R403.1.1.
c. P = greater than 2 in. but less than "T" above per R403.1.1.
d. Minimum depth = 12 in. per R403.1.4.

FIGURE 8.7

BRACING WALL FOUNDATION REQUIREMENTS FOR SDC D_0, D_1 AND D_2

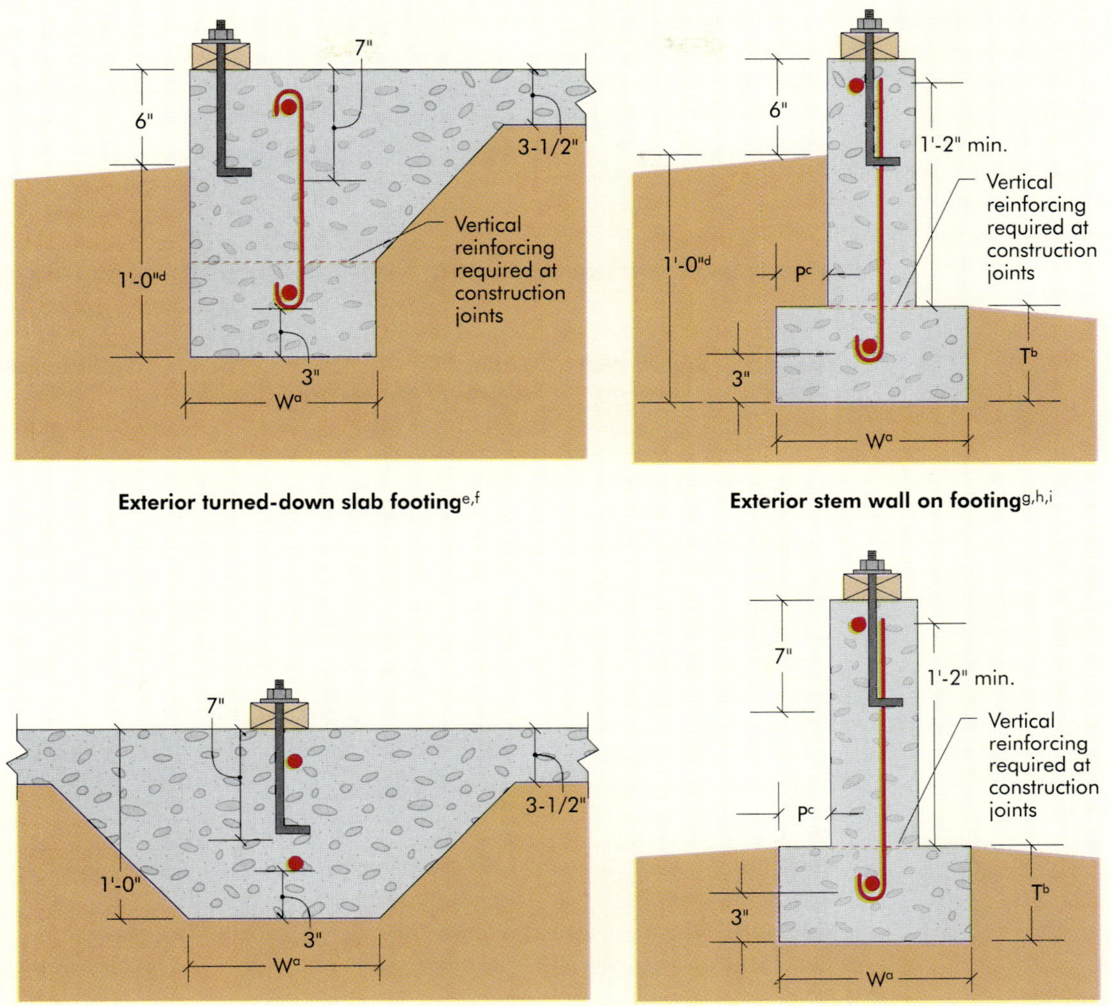

Exterior turned-down slab footing[e,f]

Exterior stem wall on footing[g,h,i]

Interior turned-down slab footing[e]

Interior stem wall on footing[g,h,i]

a. W = per Table R403.1.

b. T = 6 in. min. per R403.1.1.

c. P = greater than 2 in. but less than "T" above per R403.1.1.

d. Minimum depth = 12 in. per R403.1.4.

e. **Turned-down slab:** Reinforcing (#4 bar top and bottom) shown for SDC D_0, D_1 and D_2. If monolithic – single #5 bar or (2) #4 bars permitted in middle third per R403.1.3.2.

f. **Turned-down slab:** Vertical bars shown for SDC D_0, D_1 and D_2 when construction joint is between footing and stem wall – #4 bars at 4 ft o.c.) is recommended.

g. **Stem wall and footing:** Vertical bars shown for SDC D_0, D_1 and D_2 when construction joint is between footing and stem wall – #4 bars at 4 ft o.c.) per R403.1.3.

h. **Stem wall and footing:** Vertical bars shown for SDC D_0, D_1 and D_2 when grouted masonry stem wall over concrete footing – #4 bars at 4 ft o.c.) per R403.1.3. *While not a code requirement, similar detail for thickened slab when done in two pours is recommended.*

i. **Stem wall and footing:** Reinforcing (#4 bar top and bottom) shown for SDC D_0, D_1 and D_2 per R403.1.3.1.

Bracing wall anchorage to footings

The foundation anchorage requirements for braced wall lines are given in Chapters 4 and 6 of the IRC. Table 8.1 was developed to summarize these requirements. Note that the applicable references in the 2006 IRC are given in the table footnotes.

TABLE 8.1

Bracing wall anchorage requirements to continuous foundation

Seismic Zone	Braced Wall Location	Building Stories	Footing Type	Min. Anchor Bolt Spacing[a]	Anchorage Requirement[b]: Bolt Diameter (in.)	Anchorage Requirement[b]: Washer
A–C	Exterior	1–3	Continuous[c]	6	1/2[c]	Any
A–C	Interior	1–3	None required	6	1/2[c]	Any
C (Townhouses only)	Interior	1–3	Continuous[c]	6	1/2[c]	0.229" x 3" x 3"[f]
D_0, D_1, D_2[e]	Exterior	1–2	Continuous[c]	6	1/2[c]	0.229" x 3" x 3"[f]
D_0, D_1, D_2[e]	Exterior	3	Continuous[c]	4	1/2[c]	0.229" x 3" x 3"[f]
D_0, D_1, D_2[e]	Interior	1–3	Continuous[d]	6	1/2[c]	0.229" x 3" x 3"[f]
D_2[e]	Interior	1	Continuous[g]	6	1/2[c]	0.229" x 3" x 3"[f]
D_2[e]	Interior	2	Continuous[h]	6	1/2[c]	0.229" x 3" x 3"[f]

a. Minimum of 2 bolts required per each plate section, first bolt at each end located between 7 bolt diameters (3-1/2 in. for 1/2 in. bolt) and 12 in. of end per R403.1.6. Exceptions found in R403.1.6.

b. Bolts shall extend at least 7 in. into masonry or concrete per R403.1.6.

c. Continuous footings under all exterior bracing walls required in Section R403.1.

d. Continuous footings under all interior bracing walls in buildings with a plan dimension greater than 50 ft, per R403.1.2.

e. Applicable to townhouses in SDC C, per R403.1.6.1.

f. Per R602.11.1.

g. Continuous foundations required for interior braced wall lines not to exceed 50 ft spacing for one-story building per R602.10.9.

h. Continuous foundations at all interior braced walls required for two-story building. Some exceptions given. Refer to R602.10.9.

Winds 110 mph and greater and Seismic Design Categories E and F

Winds 110 mph and greater, and Seismic Design Categories E and F are outside the scope of the IRC and must be designed in accordance with the provisions of the International Building Code or other relevant referenced documents.

CHAPTER 9

Cripple Walls

What is a cripple wall?

A cripple wall, also called a pony wall, is a less-than-full-height wall that is used to raise the elevation of a floor above the foundation, as shown in Figure 9.1. A cripple wall can also be used in conjunction with a stepped foundation to maintain a common plate elevation as the foundation drops away from the plate line, as shown in Figure 9.2. In either case, a cripple wall is simply a stud wall that is less than full height.

FIGURE 9.1
CRIPPLE WALL USED TO RAISE FLOOR ELEVATION

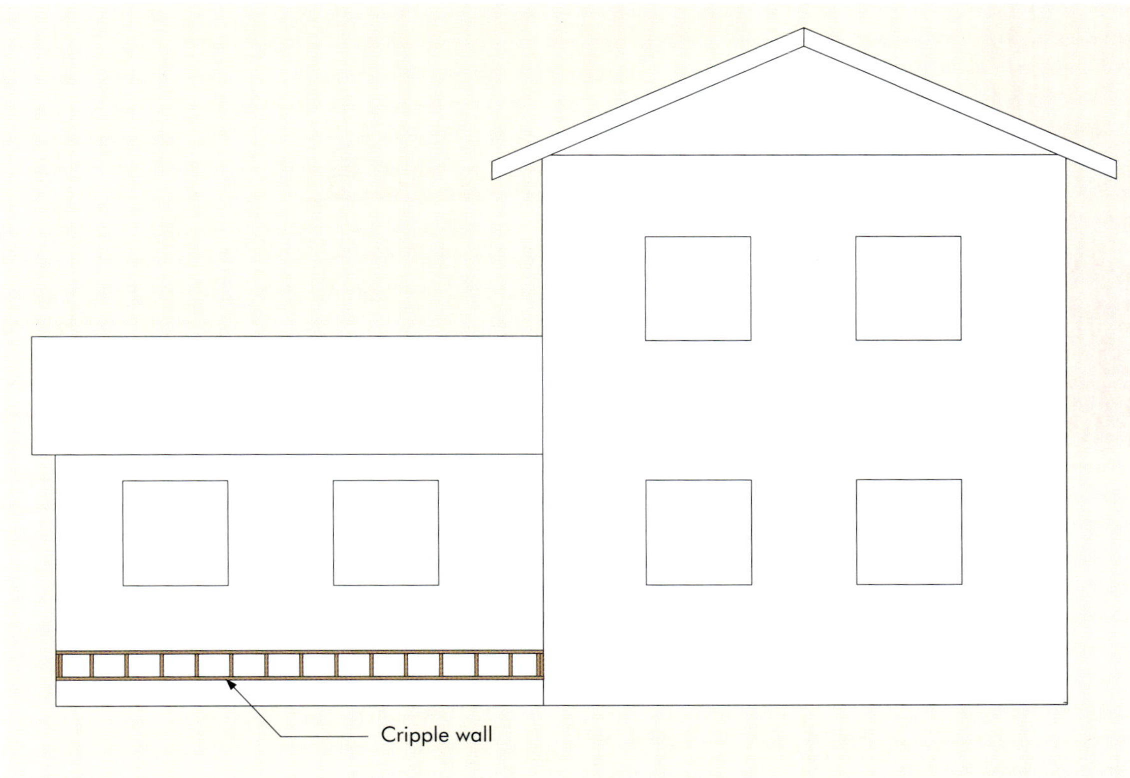

FIGURE 9.2
CRIPPLE WALL USED WITH A STEPPED FOUNDATION

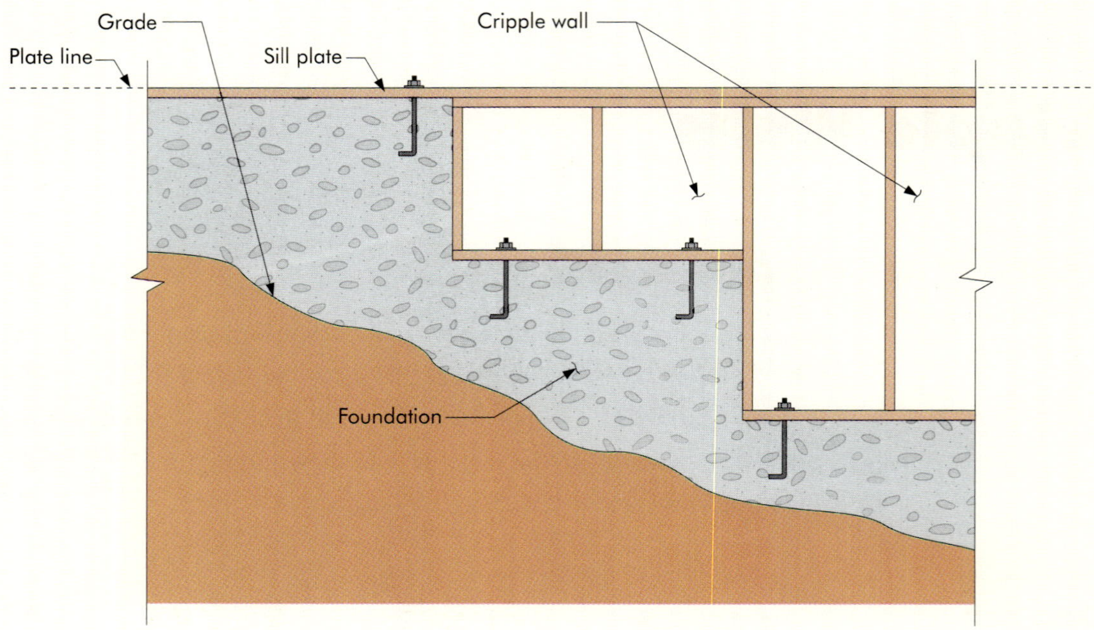

Cripple wall bracing (R602.10.2)

Even though less than full height, the cripple wall is still a part of the load path and is subject to the same kinds of loads, vertical and horizontal, as a full-height wall. In fact, because the building loads get larger as the number of stories or building height increases, the cripple wall can be expected to have even greater loads than the walls directly above. For this reason, for any seismic design category (SDC) except D_2, the IRC (Section R602.10.2.1) permits the cripple wall to be braced using any method in Table R602.10.1 as long as the amount of bracing used for the cripple wall is 15 percent greater (multiplied by a factor of 1.15). In addition, the braced wall spacing in the cripple wall should be reduced from 25 to 18 feet.

Example: A cripple wall supports the first story of a two-story residence in Bismarck, North Dakota. Table R301.2(1) of the locally adopted IRC gives a design wind speed of 90 mph and a SDC of A. The builder is planning on using structural fiberboard for bracing. How much bracing is required for the cripple wall?

Answer: From Table R602.10.1, for an SDC of A and a wind of 90 mph, first of two stories, and structural fiberboard sheathing (Method 4), 25 percent is required with a maximum spacing of 25 feet between braced wall panels. Section R602.10.2.1 requires an increase in the minimum amount of bracing by 15 percent, therefore, 25% x 1.15 = 28.8%. This section also decreases the maximum distance between braced wall panels from 25 to 18 feet.

In Seismic Design Category D_2, only Method 3 bracing is permitted for cripple walls and 75 percent of the cripple wall must be covered with wood structural panels.

Redesignation of a cripple wall into a story (R602.10.2.3)

In all wind or seismic applications, the building designer is permitted in Section R602.10.2.3 to redesignate or redefine the cripple wall as a story and then use the bracing requirements of Table R602.10.1 without the cripple wall adjustments. When redesignation is used, the cripple wall is counted as the first or ground floor wall, the first or ground floor wall becomes the second, etc.

Note that in Seismic Design Category D_2, Table R602.10.1 only permits a two-story building. This means that the redesignation option is only available for a single-story building in Seismic Design Category D_2. This is because if a two-story building is redesignated, the second story becomes a third story and the structure falls outside of the prescriptive bracing provisions of the IRC. Consequently, the bracing system of the structure would have to be engineered.

Bear in mind that redesignation is **optional**, and the builder in Seismic Design Category D_2 will have to use the Table R602.10.1 Seismic Design Category D_2 cripple wall requirements when building a two-story residence.

Cripple walls and interior braced wall lines (R602.10.9)

The IRC is not completely clear about when interior braced walls are required to have a rigid foundation in relation to cripple walls. Section R602.10.9 covers interior braced wall line support for Seismic Design Category D_2 only. It requires continuous foundations at intervals not exceeding 50 feet for single-story buildings. An exception is provided in this section that permits two-story structures in Seismic Design Category D_2 to meet the same requirements, provided all of three conditions are met. One of these three conditions is that the height of cripple walls shall not exceed 4 feet.

Cripple walls in Seismic Design Categories D_0, D_1 and D_2 (R602.10.11.4)

Obviously, as the seismic forces on a structure increase, the lateral load acting on the cripple walls also increases. In Section R602.10.9, discussed previously, provisions are made for **interior** braced wall lines placed over continuous foundations in Seismic Design Category D_2 and only at intervals not to exceed 50 feet. From this requirement we can infer that interior braced wall lines do not have to be placed over foundations in SDCs less than D_2.

In Seismic Design Categories D_0, D_1 and D_2, Section R602.10.11.4 requires increased **exterior** cripple wall bracing when interior braced walls are not supported by a continuous foundation.

This is reasonable because without direct foundation support, these interior braced wall lines transfer the lateral forces into the floor diaphragm instead. The floor diaphragm, in turn, transfers these lateral forces into the exterior foundations. When the exterior foundation wall line is supported by cripple walls, the cripple wall bracing, specified in Section R602.10.2, must be increased. This is done by multiplying the length of the required **exterior** cripple wall bracing **parallel** to the unsupported interior braced wall by a factor of 1.5, per Section R602.10.11.4.

If a Method 3 bracing (wood structural panel) is used and the length of wall bracing cannot be increased by a factor of 1.5 because of insufficient length available, the nail spacing along the perimeter of the Method 3 sheathing used for bracing shall be reduced to 4 inches o.c., per Section R602.10.11.4. There is a long history of increasing the shear capacity of wood structural panels by increasing the perimeter nailing, so it is no surprise that this option is available to Method 3 braced wall lines.

The code does not stipulate what to do if any of the other bracing methods (2, 4, 5, 7 or 8) are used and there is insufficient length within the braced wall line to permit a 50 percent increase in bracing. In this case, an engineered design or other alternative bracing approved by the building official must be incorporated. Another option might be to consider Method 3 when there is insufficient wall length to permit the 1.5 increase of the other bracing methods. Because Method 3 has smaller percentages of bracing than the other methods, the use of Method 3 for cripple wall bracing may alleviate the insufficient length problem. If not, the option to use Method 3 and decrease the perimeter nail spacing to 4 inches o.c., as permitted by R602.10.11.4, is still available.

Cripple walls and stepped foundations in Seismic Design Categories D_0, D_1 and D_2 (R602.11.3)

When a cripple wall is used in a stepped foundation in Seismic Design Categories D_0, D_1 and D_2, there are a number of additional requirements.

1. Where there is a difference in height of more than 4 feet in a given braced wall segment, the braced wall panel must be constructed according to Figure R602.11.3, reproduced in Figure 9.3.

    - When the distance "A" is greater than 8 feet in a 25-foot wall, the metal tie shown in Figure 9.3 is required.
    - When the distance "A" is less than 8 feet in a 25-foot wall, the bracing shall be applied to the cripple wall end of the wall line.

    The strap described above ties the braced portion of the wall to the unbraced portion of the wall.

    Note also that a stepped foundation shall not be constructed by placing a cripple wall over a cripple wall, as shown in Figure 9.4.

FIGURE 9.3

STEPPED FOUNDATION CONSTRUCTION

(IRC FIGURE R602.11.3)

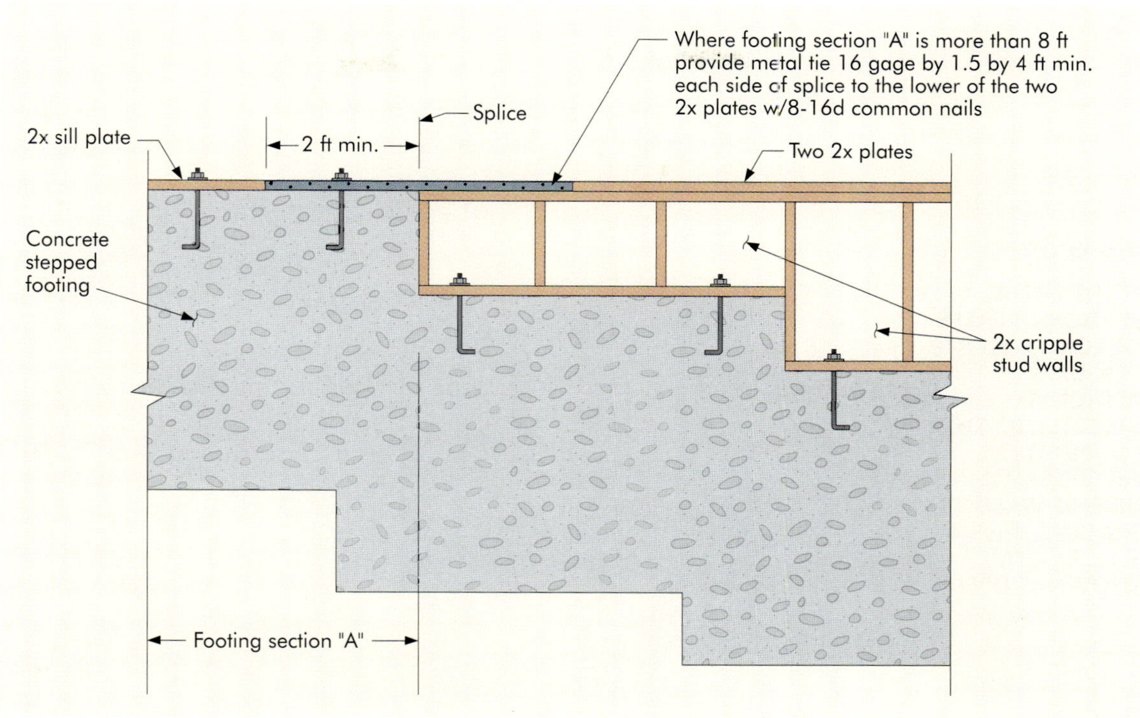

FIGURE 9.4

CRIPPLE WALLS SHALL NOT BE PLACED ONE OVER THE OTHER

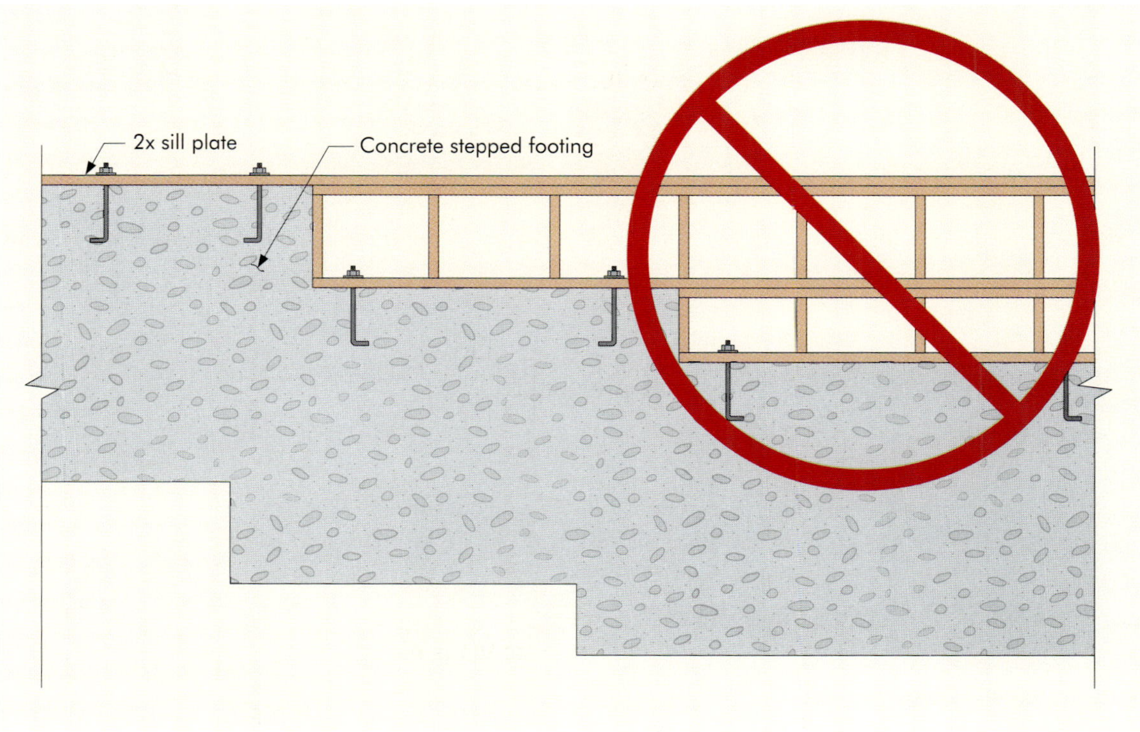

2. Section R602.11.3, item 2, further states that if 8 feet or more of the lowest floor is anchored directly to the foundation, as shown in Figure 9.5, no additional cripple wall bracing is required. In order to gain this benefit, the cripple wall top plate must extend over the foundation and be anchored with two bolts, as shown in Figure 9.5.

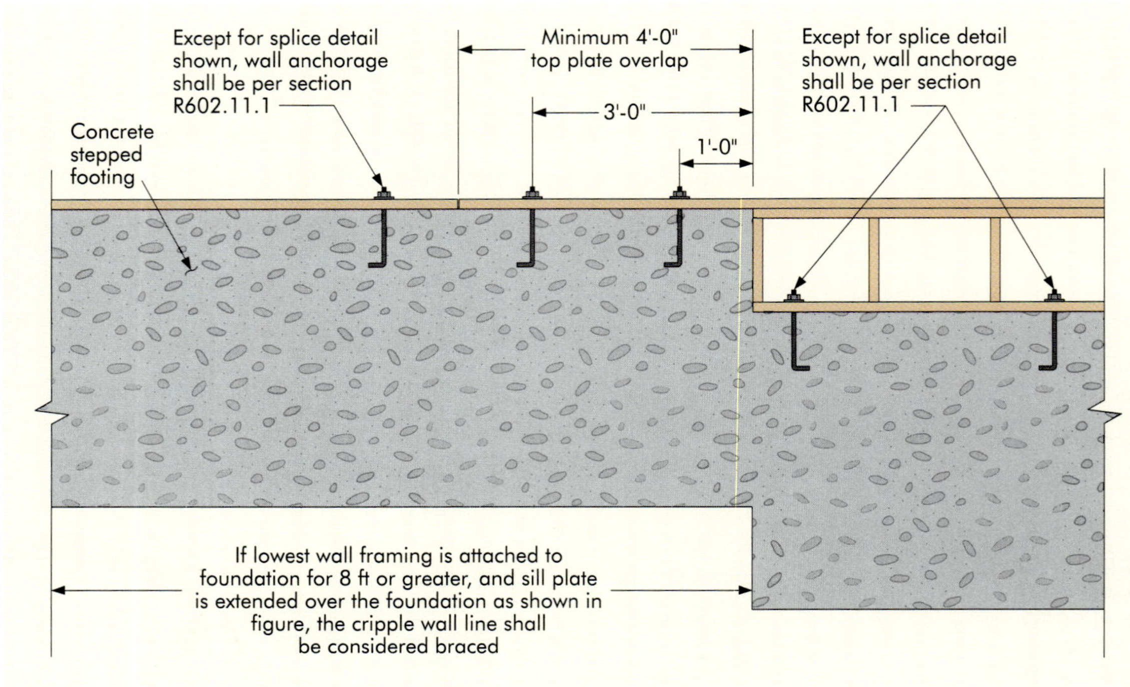

FIGURE 9.5

EIGHT LINEAL FEET OR GREATER OF LOWER-FLOOR FRAMING ATTACHED DIRECTLY TO THE FOUNDATION PROVIDES ALL CRIPPLE WALL BRACING WHEN SPLICE DETAIL SHOWN IS USED

3. (Section R602.11.3, item 3) This section clarifies that over a **stepped** footing, anywhere a cripple wall occurs in a structure (Figure 9.6), the bracing requirements of a story shall be applied to that cripple wall. If any side of a building has a cripple wall over a stepped foundation, all cripple walls above that stepped foundation have to meet the story requirements specified in R602.11.3. For Seismic Design Categories A–D_1, the requirements of R602.10.2.1 shall apply. For Seismic Design Category D_2, the cripple-wall-specific requirements of Table R602.10.1 are appropriate.

If the redesignation provisions of R602.10.2.3 are used, note that in Seismic Design Category D_2, such a foundation is limited to single-story buildings as follows: the cripple wall is designated as the first floor and the first floor becomes the second floor. In Seismic Design Category D_2, only two-story buildings are permitted (Table R602.10.1).

FIGURE 9.6
CRIPPLE WALL IN SDC D_0, D_1 AND D_2, TREATED AS A STORY

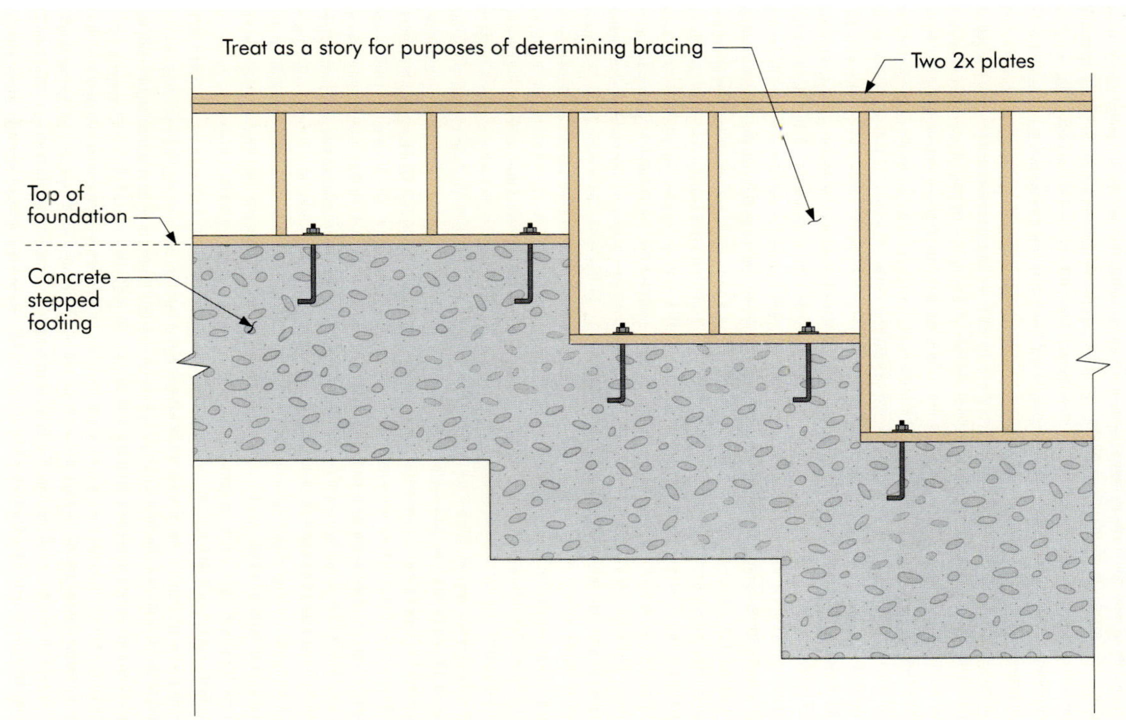

4. (Section R602.11.3 item 4) This section clarifies that when only the bottom or footing of a foundation is stepped and the top of the footing has a common elevation, this situation is not considered, for purposes of the IRC, a stepped footing. The normal attachment requirements in Section R602.11.1 apply. See Figure 9.7.

FIGURE 9.7
FOR PURPOSES OF INTERPRETING THE IRC, THIS IS NOT A STEPPED FOOTING

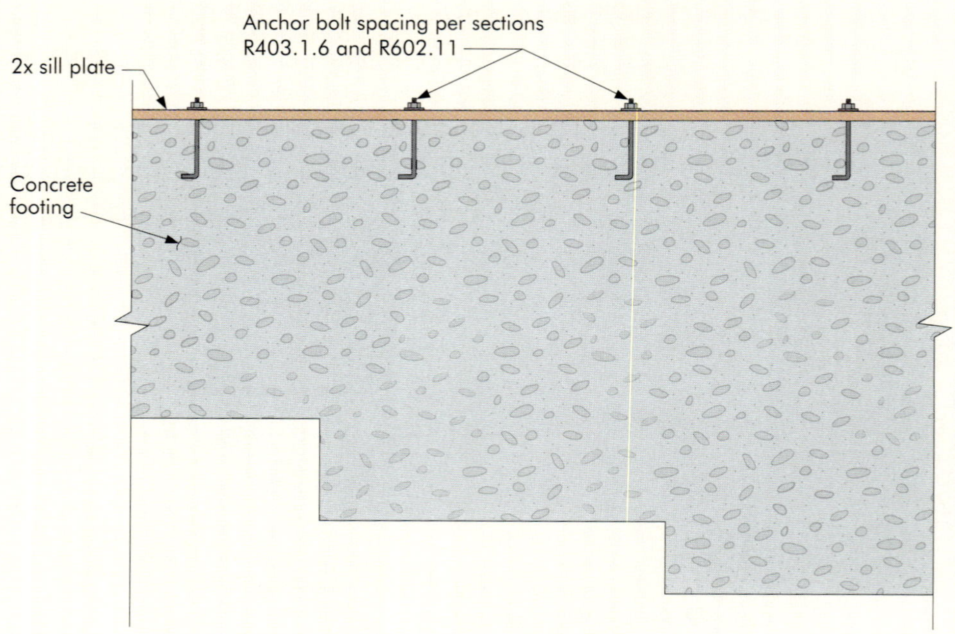

Summary of cripple wall provisions

Table 9.1 shows a summary of the most commonly used cripple wall bracing provisions.

TABLE 9.1

Cripple wall bracing

<table>
<tr><th>Case</th><th>SDC</th><th>Cripple Wall Bracing</th></tr>
<tr><td rowspan="2">1</td><td>$A–D_1$</td><td>Brace same method as wall above but add 15% to bracing required by Table R602.10.1 and reduce spacing of braced wall panel to 18 ft on center
or
Redesignate cripple wall as first story, per R602.10.2.3, and brace all walls accordingly</td></tr>
<tr><td>D_2</td><td>Brace as required for cripple wall, as shown in Table R602.10.1
or
Redesignate cripple wall as first story, per R602.10.2.3, and brace all walls accordingly</td></tr>
<tr><td rowspan="2">2</td><td>$A–D_1$</td><td>Brace same as wall above but add 15% to bracing required by Table R602.10.1 and reduce spacing of braced wall panel to 18 ft on center
or
Redesignate cripple wall as first story, per R602.10.2.3, and brace all walls accordingly</td></tr>
<tr><td>D_2</td><td>Brace as required for cripple wall, as shown in Table R602.10.1</td></tr>
<tr><td rowspan="2">3</td><td>$A–D_1$</td><td>Brace same as wall above but add 15% to bracing required by Table R602.10.1 and reduce spacing of braced wall panel to 18 ft on center</td></tr>
<tr><td>D_2</td><td>Not permitted, per Section R301.2.2.4.1, which refers to Table R602.10.2.3 for story limitations</td></tr>
</table>

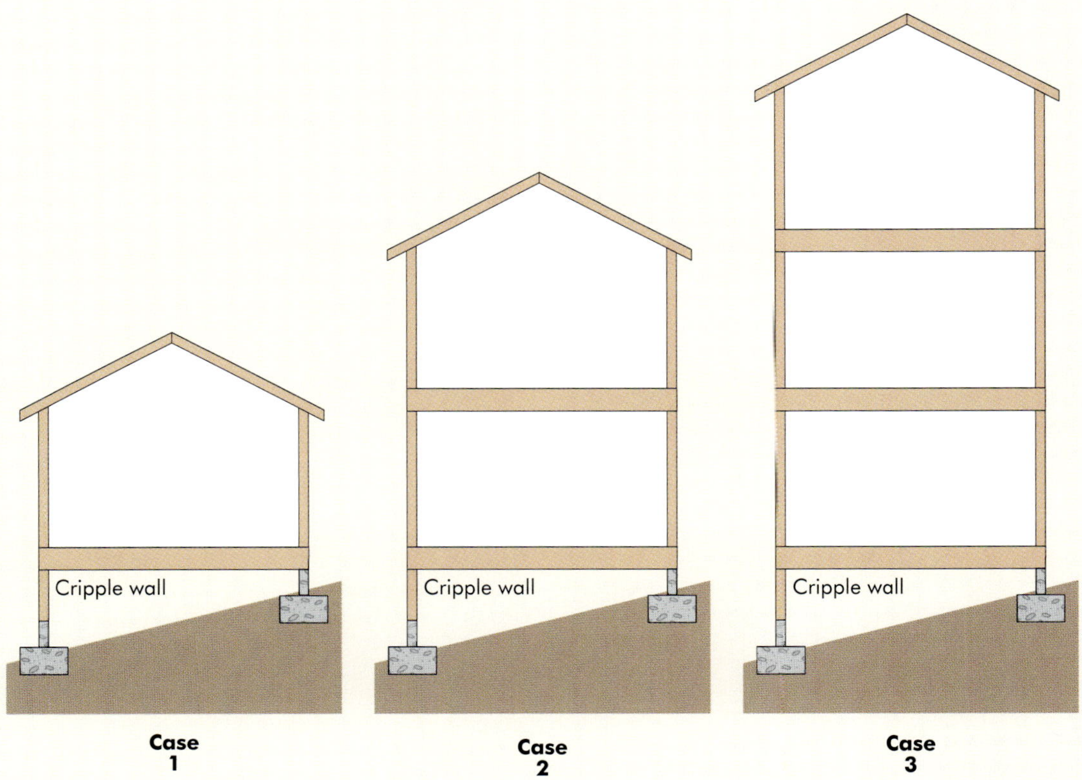

CHAPTER 10

Whole House Considerations

Putting it all together

Previous chapters in this book have explained the individual bracing requirements separately. This chapter puts together all the individual IRC requirements into examples that consider the whole house. Examples will be used to demonstrate wall bracing for the whole house using modern house plans. Symbols will be used, as shown in Table 10.1, to call out which bracing type is used in each example. Detailed requirements for each bracing type are provided in Chapter 5.

What's the intent?

In situations not specifically covered by the IRC – gray areas – a couple of good questions to ask are what is the intent of the specific provision, and is the intent being met? In these situations, it is essential to communicate with the local building official to determine their interpretation. In many cases, they may have already encountered the problem and local requirements and solutions may already be in place.

Any unusual bracing issues should be discussed with the building official before the plans are finalized to avoid extensive redesign.

TABLE 10.1

Bracing schedule used to describe the bracing type used in Chapter 10 examples

Symbol	Reference	Description	Minimum Length	Note
1	R602.10.3 Method 1	1" x 4" diagonal let-in bracing	Varies, 60 to 45 degrees from horizontal	8' wall = 4'-7" to 8' 9' wall = 5'-2" to 9' 10' wall = 5'-9" to 10'
2	R602.10.3 Method 2	Diagonal wood boards	4'	Attach per Table R602.3(1)
3	R602.10.3 Method 3	Wood structural panel	4'	Attach per Tables R602.3(3) and R602.3(1), (2)
4	R602.10.3 Method 4	Structural fiberboard	4'	Attach per Table R602.3(1)
5	R602.10.3 Method 5	Gypsum wallboard	8' one side, 4' both sides	Attach per Table R602.3(1) for sheathing, Table R702.3.5 for interior gypsum board
6	R602.10.3 Method 6	Particleboard	4'	Attach per Tables R602.3(1), (4)
7	R602.10.3 Method 7	Portland cement plaster on studs	4'	Attach per Section 703.6
8	R602.10.3 Method 8	Hardboard panel siding	4'	Attach per Table 703.4
A1	R602.10.6.1	Bracing panel with hold downs	28" one side – no floor above 28" two sides – one floor above	With hold downs. Min. length varies 28" to 42"
A2	R602.10.6.2	Portal frame with hold downs	16" – no floor above 24" – one floor above	With hold downs
C	R602.10.5	Wood structural panel, continuously sheathed	24" and up	All exterior surfaces sheathed except openings
C.b	Table R602.10.5, Footnote b	Wood structural panel, continuously sheathed	8' wall = 24" 9' wall = 27" 10' wall = 30"	Adjacent to garage door opening, light roof, single story, 4:1 aspect ratio
C.c	Table R602.10.5, Footnote c	Portal frame without hold downs, continuously sheathed	8' wall = 16" 9' wall = 18" 10' wall = 20"	Adjacent to garage door opening, first of two-story, 6:1 aspect ratio

About design wind speeds

- Wind speeds 110 mph and greater are beyond the scope of the IRC per Section R301.2.1.1.

- Wind speeds 100 mph and greater in hurricane-prone regions are beyond the scope of the IRC per Section R301.2.1.1. Hurricane-prone regions are defined in R202 as:

> *Areas vulnerable to hurricanes, defined as the U.S. Atlantic Ocean and Gulf of Mexico coasts where the basic wind speed is greater than 90 miles per hour (145 km/h), and Hawaii, Puerto Rico, Guam, Virgin Islands, and American Samoa.*

- Special wind regions also exist and are shown in IRC Figure R301.2(4). Design wind speeds in special wind regions are not defined in the IRC, but represent any unusual wind conditions, such as mountainous terrain, gorges, or other conditions that may result in higher local wind speeds. Refer to ASCE7-05 Section 6.5.4.1 for a definition. If unsure about the design wind speed for a given location, check with the authority having jurisdiction.

- Wall bracing requirements, per IRC Table R602.10.1, can be divided into two different categories for wind speeds:

1) "100 mph and less." This means that for wall bracing purposes, in the category of "100 mph and less" wind speeds that are 85 to 100 mph have the exact same wall bracing requirements per Table R602.10.1.

2) "greater than 100 mph, but less than 110 mph." In the category of "greater than 100 mph, but less than 110 mph" wind speeds that are 100 to 110 mph also have the exact same wall bracing requirements per Table R602.10.1.

Example 10.1 shows how a simple house can comply with the IRC wall bracing provisions. Later examples will show more complex designs.

Example 10.1: A simple house in Seismic Design Category (SDC) C and wind zone 90 mph. Isolated 4-foot structural fiberboard bracing segments (Method 4) will be used to brace the house.

Figures 10.1a and 10.1b show the house and braced wall segments for each story. Table 10.2 summarizes the amount of bracing required and the amount of bracing provided for each braced wall line on each story.

FIGURE 10.1a
FIRST STORY FLOOR PLAN WITH METHOD 4 (STRUCTURAL FIBERBOARD) BRACING SEGMENTS SHOWN

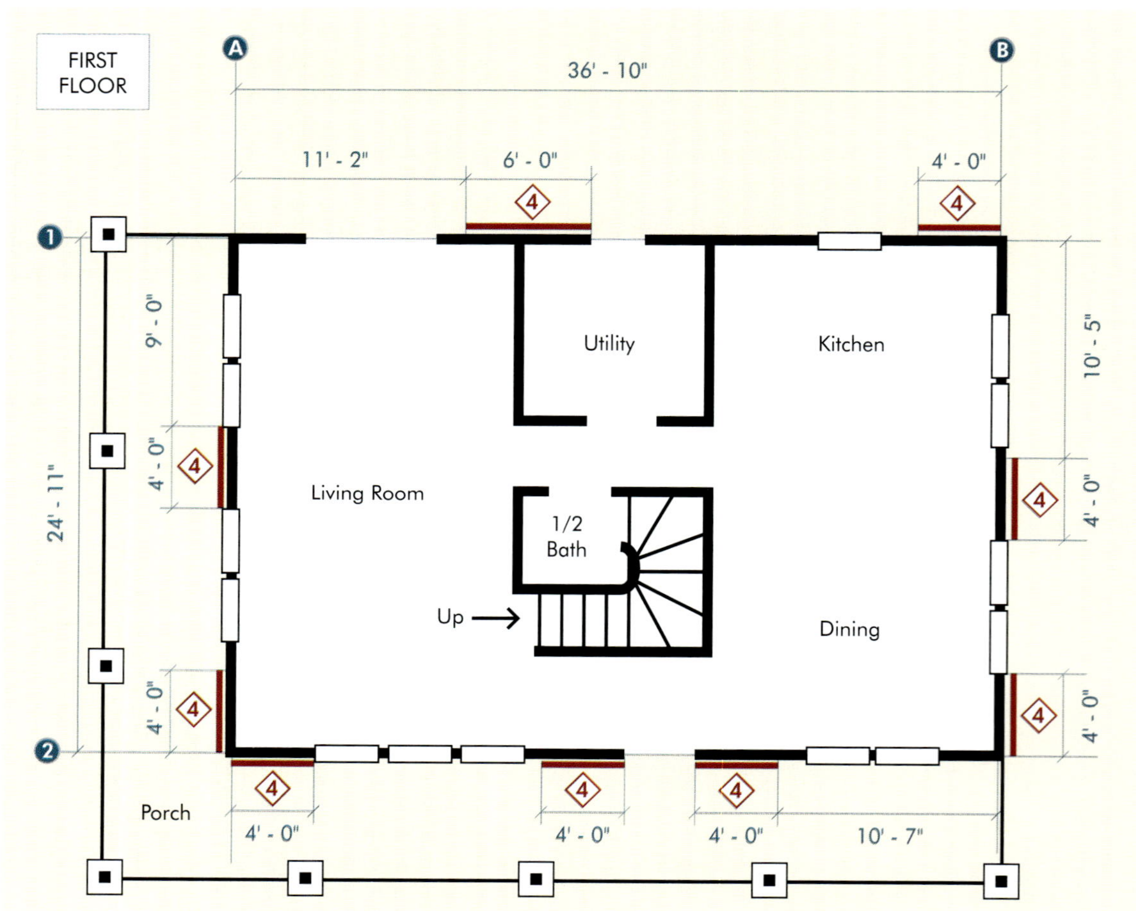

This example highlights several important issues:

- The bracing amount required is based on the "100 mph or less" wind speed category because detached one- and two-family dwellings in Seismic Design Category C are exempt from seismic requirements (R301.2.2, Exception).
- In Seismic Design Categories A, B and C for detailed one- and two-family dwellings, wall bracing on the second floor can be placed over an opening on the first floor. This is seen along lines 1 and 2 in Figures 10.1a and 10.1b.
- Bracing segments are permitted to occur up to 12.5 feet away from the ends of braced wall lines (R602.10.1).
- How to accommodate a braced wall line spacing greater than 35 feet.
- How to compare the amount of bracing required to the amount of bracing present.

TABLE 10.2

BRACING REQUIREMENT FOR EXAMPLE 10.1

The amount of bracing required and the amount of bracing provided for each braced wall line on each story for Example 10.1

Story	Braced Wall Line	Bracing Method	Bracing Amount Required %	Bracing Length Provided (ft)	Braced Wall Line Length (ft)	Bracing Amount Provided %	Status	Notes
First of Two	1	4	25%	10	36.83	27%	OK	--
	2	4	25%	12	36.83	33%	OK	
	A	4	26.3%	8	24.92	32%	OK	1
	B	4	26.3%	8	24.92	32%	OK	
Top of Two	1	4	16%	8	36.83	22%	OK	--
	2	4	16%	8	36.83	22%	OK	
	A	4	16.8%	8	24.92	32%	OK	1
	B	4	16.8%	8	24.92	32%	OK	

Note:

1. For wall lines A and B, since braced wall line spacing exceeds 35 ft, the bracing amount required must increase by a factor of 36.83/35 = 1.05. (From R602.10.1.1. Note that requirements of both exceptions are met in this example.)

FIGURE 10.1b

SECOND STORY FLOOR PLAN WITH METHOD 4 BRACING SEGMENTS SHOWN

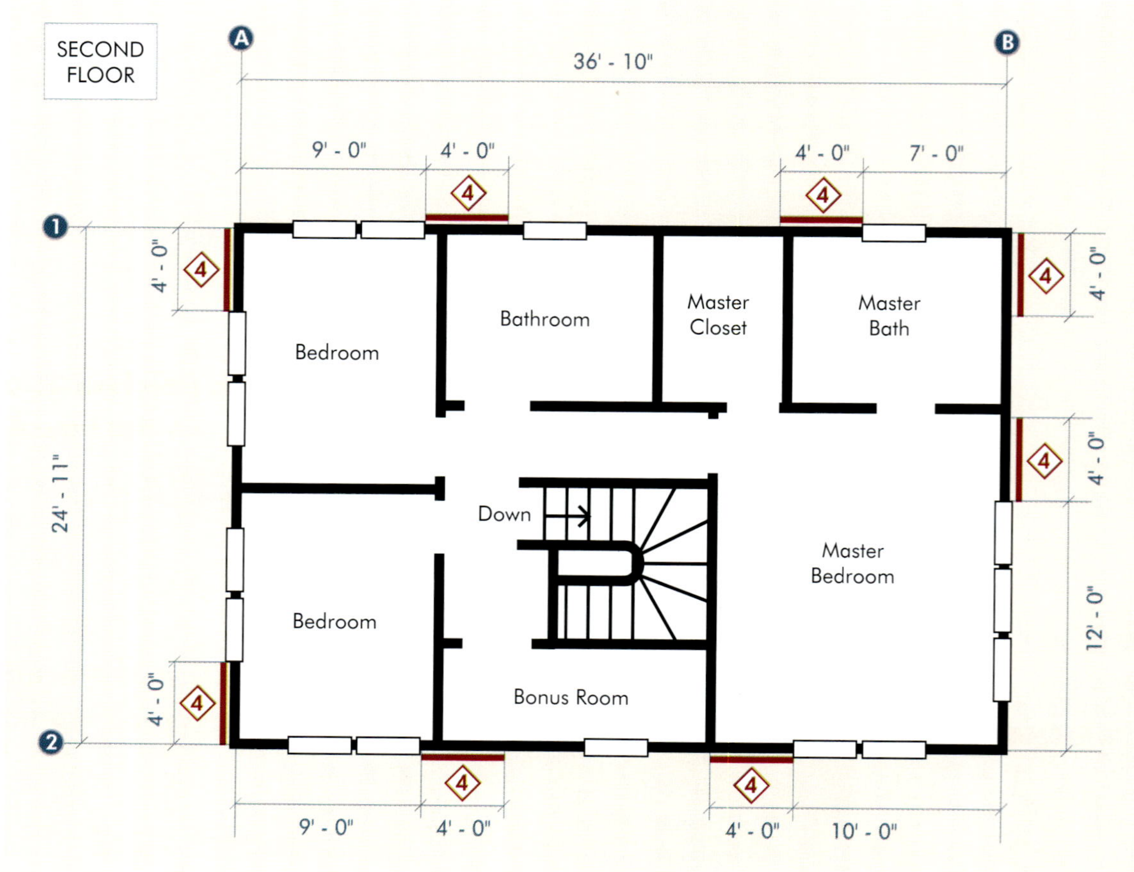

Although Method 4 (structural fiberboard) bracing was used in Example 10.1, the solution would be identical if Methods 2, 6, 7 or 8 were used. Although the required length of wall bracing is less for Method 3 bracing, Method 3 could also be used for these floor plans.

Note that a number of the bracing panels shown above could have been placed at the corners, but were not in this example to illustrate the provisions of R602.10.1 that permit the panels to be placed away from the corners up to 12-1/2 feet. From an engineering perspective, corner placement is always preferred but given the redundancies and loads that can be expected on residential-sized buildings, displacement away from corners is permitted in the prescriptive requirements.

Are Methods 1 and 5 appropriate for this example?

With extensive reworking of the plans, Method 1 (let-in bracing) could probably be made to work. Method 5 (gypsum wallboard or sheathing) could be used, although certain exterior walls would have to have exterior gypsum sheathing on the outside and interior gypsum wallboard on the inside, each with its own fastener schedule to meet the 4-foot length requirement of Section R602.10.4.

Example 10.2: A simple house in Seismic Design Category D_2 and wind zone 85 mph. Isolated wood structural panel bracing (Method 3), and gypsum bracing segments (Method 5) will be used to brace the house.

Figures 10.2a and 10.2b show the house and braced wall segments for each story. Table 10.3 summarizes the amount of bracing required and the amount of bracing provided for each braced wall line on each story.

FIGURE 10.2a

FIRST STORY FLOOR PLAN WITH METHOD 3 (WOOD STRUCTURAL PANELS) AND METHOD 5 (GYPSUM) BRACING SEGMENTS SHOWN

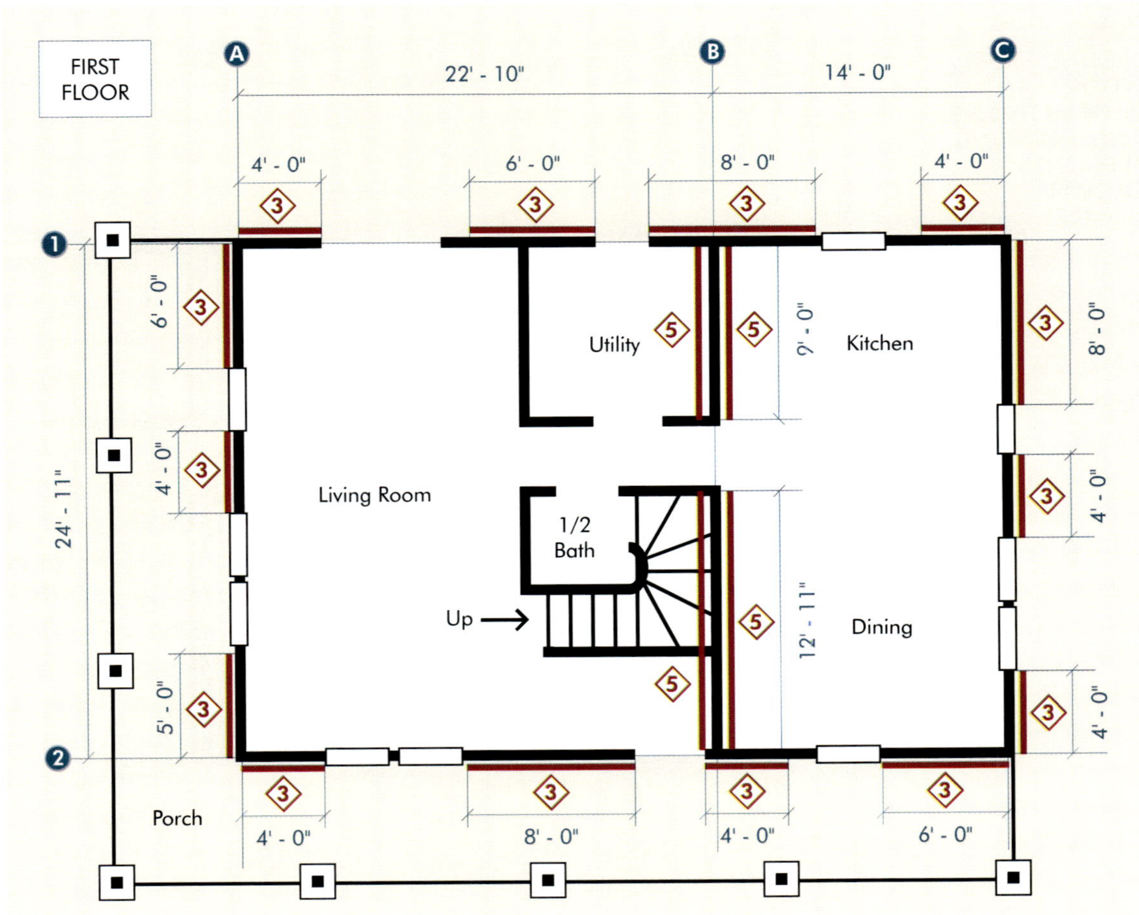

Example 10.2 highlights several important issues:

- The bracing amount required is based on the Seismic Design Category D_2 requirements.
- In Seismic Design Categories C, D_0, D_1 and D_2, wall bracing on the second floor cannot be placed over an opening on the first floor except in certain cases (R301.2.2.2.2, Item 3, Exceptions).
- In Seismic Design Categories D_0, D_1 and D_2 ,bracing segments are not permitted to occur away from the ends of braced wall lines except in certain cases and only with Method 3 (R602.10.11.2, Exception).
- Braced wall line spacing is limited to 25 feet (R602.10.11.1).
- How to compare the amount of bracing required to the amount of bracing present.

FIGURE 10.2b
SECOND STORY FLOOR PLAN WITH METHOD 3 AND 5 BRACING SEGMENTS SHOWN

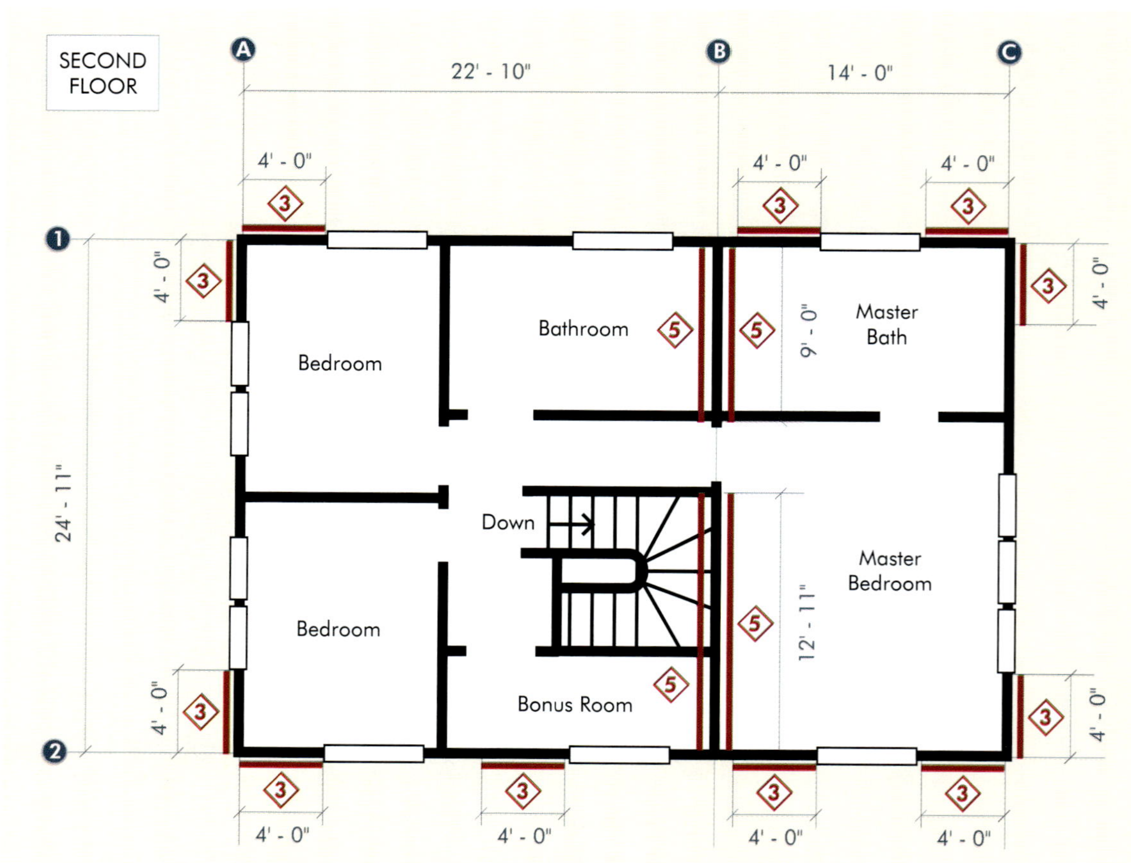

TABLE 10.3

BRACING REQUIREMENT FOR EXAMPLE 10.2

The amount of bracing required and the amount of bracing provided for each braced wall line on each story for Example 10.2

Story	Braced Wall Line	Bracing Method	Bracing Amount Required %	Bracing Length Provided (ft)	Braced Wall Line Length (ft)	Bracing Amount Provided %	Status	Notes
First of Two	1	3	55%	22	36.83	60%	OK	--
	2	3	55%	22	36.83	60%	OK	
	A	3	55%	15	24.92	60%	OK	--
	B	5	75%	21.92	24.92	88%	OK	
	C	3	55%	16	24.92	64%	OK	
Top of Two	1	3	25%	12	36.83	33%	OK	--
	2	3	25%	12	36.83	33%	OK	
	A	3	25%	8	24.92	32%	OK	--
	B	5	40%	21.92	24.92	88%	OK	
	C	3	25%	8	24.92	32%	OK	

Code "gray areas"

Before we look at Example 10.3, a more complex house plan bracing example, let's consider some of the commonly occurring complexities – "gray areas" – and possible solutions.

What to do when exterior wall lines have an offset that exceeds 4 feet

It is relatively easy to apply the bracing provisions to a simple rectangular box-like house plan; however, when house plans become more complex, as is typical with modern house plans which include large offsets, "L" shapes and other irregularities, applying the bracing provisions can be a challenge. One of the first steps to wall bracing is to define the braced wall lines. This can become tricky when walls have offsets that exceed 4 feet, as discussed in Chapter 7 and shown in Figure 10.3. Although the intent of the IRC is that bracing wall lines begin and end at opposite exterior walls, this can be difficult to apply to modern house plans that are not simple rectangular boxes. Figures 10.4 to 10.8 provide potential known solutions to the issues presented in Figure 10.3. In many modern house plans, the intent of the code, with respect to braced wall line start and stop locations, falls into a gray area and becomes subject to interpretation.

FIGURE 10.3

WHAT TO DO WHEN *EXTERIOR* WALL LINES HAVE AN OFFSET THAT EXCEEDS 4 FEET

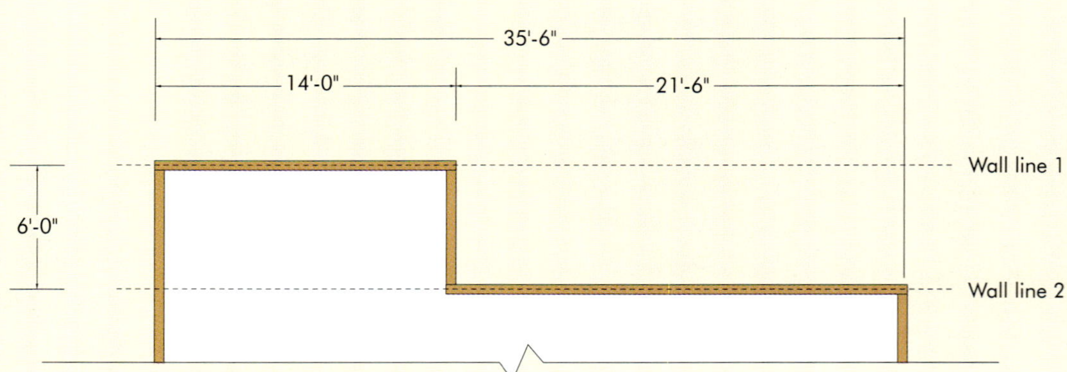

Issues:
1. Wall lines 1 and 2 have an offset that exceeds 4 ft.
2. If wall line 2 is treated as a braced wall line, where does it start and stop?

Figure 10.3 shows a common feature in many house designs – offsets that exceed 4 feet. When such a feature occurs, there are quite a few ways to provide the required bracing.

The preferred solution is to make slight modifications to the design, as shown in Figures 10.4 and 10.5, so that the offset can be modified such that the house complies with a literal interpretation of the IRC bracing provisions.

FIGURE 10.4

MODIFYING THE WALL LINES SO THAT THE OFFSET COMPLIES WITH THE 4-FOOT LIMIT

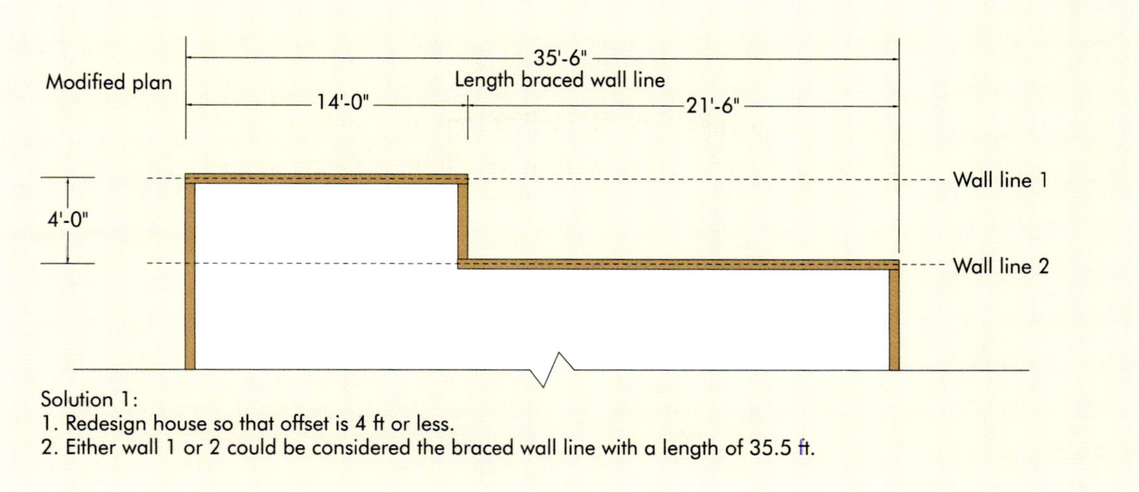

FIGURE 10.5

MODIFYING THE WALL LINES SO THAT AN ACTUAL BRACED WALL PANEL OCCURS IN THE DEFINED BRACED WALL LINE

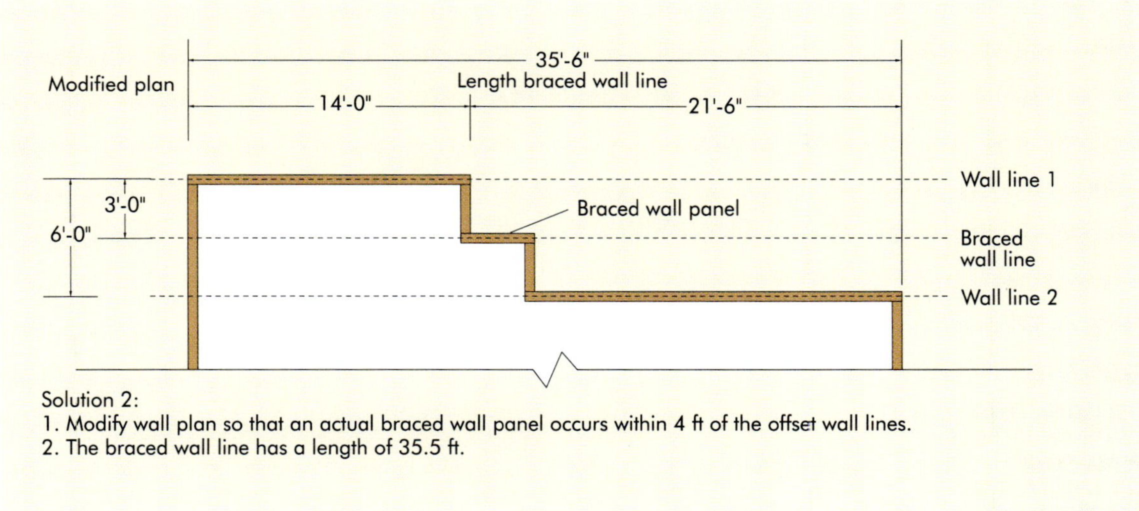

Figure 10.6 illustrates the requirements that, if wall line 2 is defined as a braced wall line, then it is 35.5 feet long. This option is certainly conservative, as it requires a designed collector (engineering). Furthermore, defining braced wall line 2 to be 35.5 feet requires more bracing on the existing wall section in order to meet the required percentage.

FIGURE 10.6

DEFINING BRACED WALL LINE 2 AS 35.5 FEET LONG

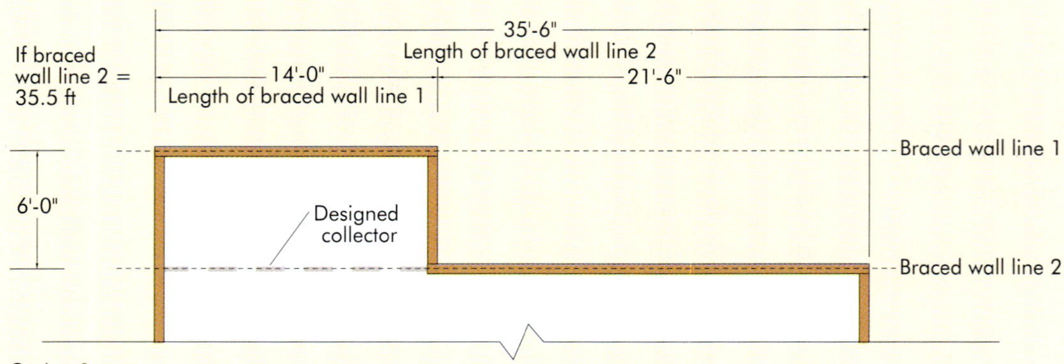

Option 1:
1. Consider wall lines 1 and 2 separate braced wall lines.
2. Braced wall line 1 has a length of 14 ft, and braced wall line 2 has a length of 35.5 ft.
3. Since a braced wall panel cannot occur within 12.5 ft from the end of braced wall line 2, a designed collector is needed.

Another potential option is shown in Figure 10.7, where braced wall line 2 is defined to be 21.5 feet long. This solution uses a "box" theory, where different sections of the house are defined as boxes. This concept is not in the IRC, but does occur in the 2006 IBC bracing provisions, per Figure 2308.9.3, where parallel braced wall lines "need not align as long as each has a braced wall panel at each end." The box theory can be challenging because defining the boxes is open to interpretation and deciding what the requirements are for each box is not defined. In addition, there may be added complexity when boxes overlap. An example of such a complexity would be that of a common wall line between the two boxes. The wall line would have to meet the bracing requirements of ***both*** the left ***and*** right boxes.

FIGURE 10.7

DEFINING THE ENDS OF BRACED WALL LINES SO THAT EACH BRACED WALL LINE REQUIRES A BRACED WALL PANEL AT EACH BRACED WALL LINE END

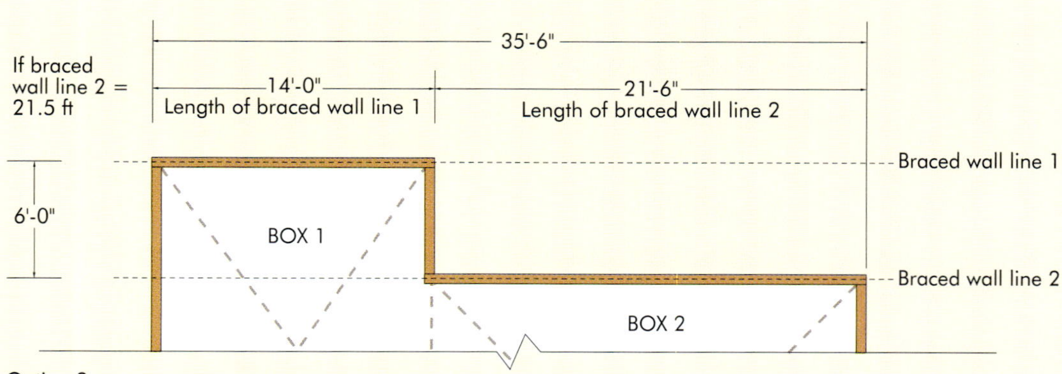

Option 2:
1. Consider wall lines 1 and 2 separate braced wall lines.
2. Braced wall line 1 has a length of 14 ft, and braced wall line 2 has a length of 21.6 ft.

Yet another potential option is the use of an "imaginary" braced wall line, as discussed in Chapter 7 and shown in Figure 10.8. This concept is not in the IRC and the IRC explicitly states in Section R602.10.1 that "braced wall lines shall consist of braced wall panel construction…"; however, the concept has some support and is known to be acceptable to some building officials. It is important to note that when using the imaginary braced wall line concept, the distance between braced wall lines is measured from the location of the imaginary wall line. Using this method, the length of the imaginary braced wall line is 35 feet 6 inches for converting percent bracing to actual length of the braced wall line. The final length of the braced wall line must be distributed between wall lines 1 and 2.

FIGURE 10.8

USING AN IMAGINARY BRACED WALL LINE

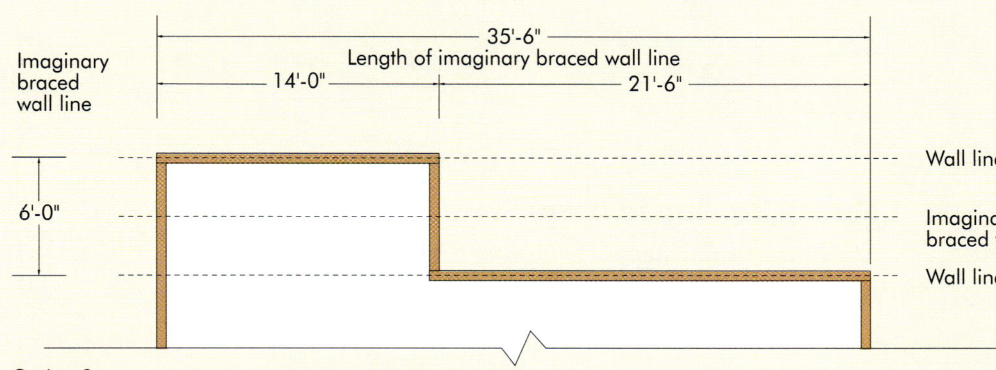

Option 3:
1. Create an "imaginary" braced wall line between wall lines 1 and 2 so that offsets do not exceed 4 ft.
2. The imaginary braced wall line has a length of 35.5 ft.
3. Distribute bracing into both wall lines.

What to do when *interior* wall lines have an offset that exceeds 4 feet

Offset interior braced wall lines pose similar problems as offset exterior braced wall lines, as shown in Figure 10.9. The issue of defining the length of the braced wall line (necessary for computing percent bracing) is a challenge. For the case shown in Figure 10.9, solutions could be:

1. Treat the interior wall lines as separate wall lines, designate one as the braced wall line and design a collector for this braced wall line since it does not extend all the way across the structure. This would be conservative because an engineered collector is provided. On the other hand, because the amount of bracing required is based on the braced wall line length equal to 33 feet, the interior braced wall may have to be lengthened to provide sufficient bracing.

2. Find the portion of the structure that does not meet the bracing requirements and either: a) Re-design the non-compliant portion, or b) engineer that portion of the structure in accordance with the IBC or other approved design documents.

FIGURE10.9

WHAT TO DO WHEN INTERIOR WALL LINES HAVE AN OFFSET THAT EXCEEDS 4 FEET

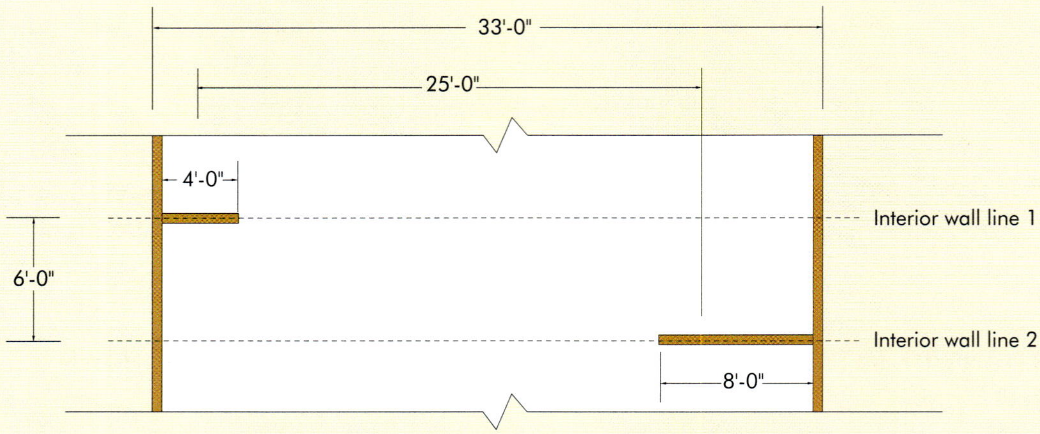

Issues:
1. Interior wall lines 1 and 2 have an offset that exceeds 4 ft.
2. If wall lines 1 and 2 are treated as separate braced wall lines, they require a collector, or a bracing panel at each end.

Mixing bracing in a wall line

As discussed in Chapter 5, mixing different bracing methods within a single braced wall line is permitted. The code does not include provisions, however, for the amount of bracing a mixed bracing type wall line is required to have. In the absence of specific provisions, it is recommended that the required bracing amount for the braced wall lines with mixed sheathing types be based on the higher amount required of the multiple bracing types. For example, if a single braced wall line has mixed Method 3 and Method 5 bracing, the total amount of bracing required for that braced wall line should be based on Method 5 requirements because they are higher than the Method 3 amount of bracing requirements.

Back to the examples...

Several of these gray areas occur in the following Example 10.3. This example has large wall offsets and mixed bracing types within one braced wall line.

Example 10.3: A one-story house in Seismic Design Category A and wind zone 95 mph. Four-foot wood structural panel bracing segments (Method 3), alternate bracing segments (R602.10.6.1 and R602.10.6.2), and gypsum bracing segments (Method 5) will be used to brace the house.

Figure 10.10 shows the house and braced wall segments. Table 10.4 summarizes the amount of bracing required and the amount of bracing provided for each braced wall line. In Figure 10.10, the braced wall lines for this design are marked in red and designated.

FIGURE 10.10

FLOOR PLAN WITH METHOD 3, METHOD 5 AND ALTERNATE BRACING SEGMENTS SHOWN

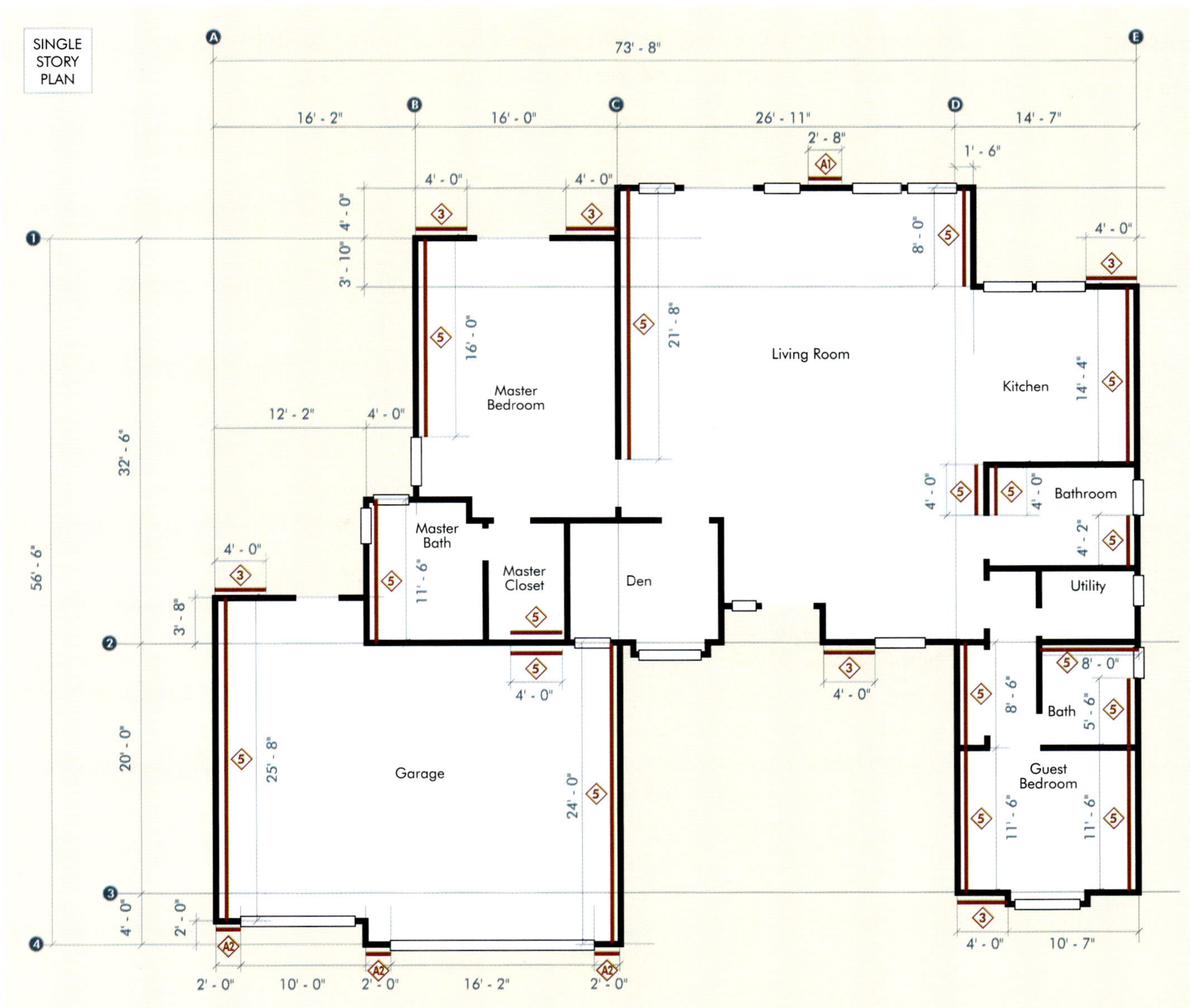

Example 10.3 highlights several important issues:

- The bracing amount required is based on 95 mph winds, which is the "100 mph or less" category.
- Use of the 24-inch-wide (A2) portal frame alternate bracing (R602.10.6.2).
- Use of the 32-inch-wide (A1) alternate bracing (R602.10.6.1).
- Use of different bracing methods in one wall line.
- Defines the length of braced wall lines "B" and "3" using the "box" concept, as discussed previously and shown in Figure 10.7.
- Describes how to check the amount of bracing required to the amount of bracing present.

TABLE 10.4

BRACING REQUIREMENT FOR EXAMPLE 10.3

The amount of bracing required and the amount of bracing provided for each braced wall line for Example 10.3

Story	Braced Wall Line	Bracing Method	Bracing Amount Required %	Bracing Length Provided (ft)	Braced Wall Line Length (ft)	Bracing Amount Provided %	Status	Notes
One Story	1	3, A1	16%	16	57.5	28%	OK	1
	2	3,5	16%	20	73.67	27%	OK	2
	3	3	16%	4	14.58	27%	OK	3
	4	A2	16%	12	32.17	37%	OK	4
	A	5	16%	25.67	25.67	100%	OK	5
	B	5	16%	27.5	32.5	85%	OK	6
	C	5	16%	45.67	60.5	75%	OK	7
	D	5	16%	32	56.5	57%	OK	8
	E	5	16%	35.5	48.67	73%	OK	9

Notes:

1. Braced wall line 1 contains bracing segments with Methods (reading left to right): 3, 3, A1 and 3.
2. Braced wall line 2 contains bracing segments with Methods (reading left to right): 3, 5 (double sided), 3 and 5. For braced wall line 2 with different bracing types in the same wall, the amount of bracing required is based on the most stringent requirements of the multiple types of bracing used in the same braced wall line.
3. Braced wall line 3 contains a single bracing segment of Method 3.
4. Braced wall line 4 contains bracing segments with Methods (reading left to right): A2, A2 and A2.
5. Braced wall line A contains a single bracing segment of Method 5.
6. Braced wall line B contains two bracing segments, both Method 5. Braced wall line B ends at the garage.
7. Braced wall line C contains two bracing segments, both Method 5.
8. Braced wall line D contains bracing segments with Methods (reading top to bottom): 5, 5 (double-sided), 5 and 5.
9. Braced wall line E contains three Method 5 bracing segments.

Example 10.4: A one-story house in Seismic Design Category A and wind zone 85 mph. Let-in (Method 1), diagonal wood boards (Method 2), gypsum (Method 5), particleboard (Method 6), portland cement plaster (Method 7) and hardboard panel siding (Method 8) will be used to brace the house.

Figure 10.11 shows the house and braced wall segments. Table 10.5 summarizes the amount of bracing required and the amount of bracing provided for each braced wall line. In Figure 10.11, the braced wall lines for this design are marked in red and designated.

FIGURE 10.11

FLOOR PLAN WITH METHOD 1, 2, 5, 6, 7 AND 8 BRACING SEGMENTS SHOWN

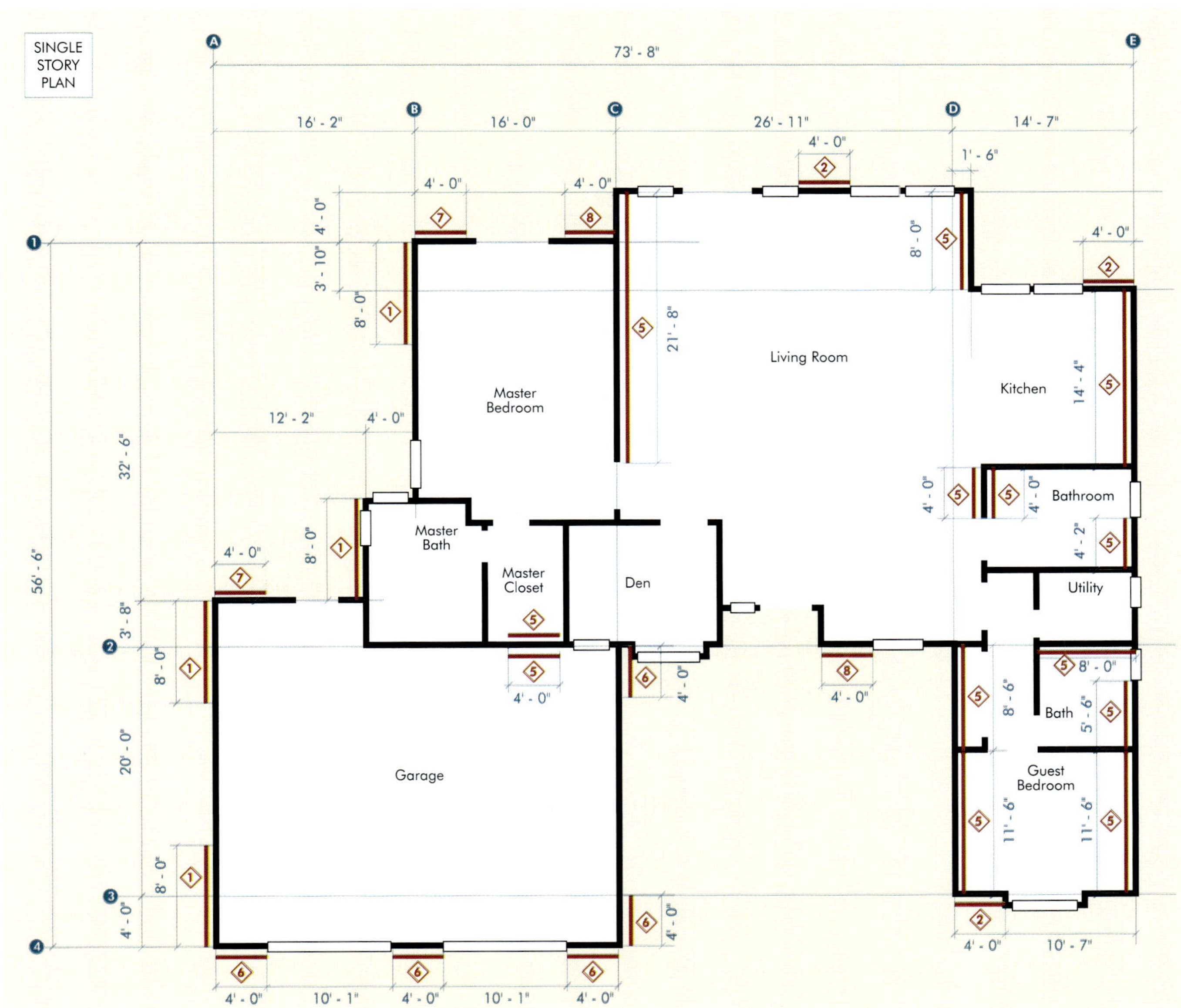

Example 10.4 highlights several important issues:

- The bracing amount required is based on 85 mph winds.
- Use of let-in bracing.
- Use of different bracing methods in one wall line, and on different wall lines.
- Defines the length of braced wall lines "B" and "3" using the "box" concept, as discussed previously and shown in Figure 10.7.
- Describes how to check the amount of bracing required to the amount of bracing present.

TABLE 10.5

BRACING REQUIREMENT FOR EXAMPLE 10.4

The amount of bracing required and the amount of bracing provided for each braced wall line for Example 10.4

Story	Braced Wall Line	Bracing Method	Bracing Amount Required %	Bracing Length Provided (ft)	Braced Wall Line Length (ft)	Bracing Amount Provided %	Status	Notes
One Story	1	2, 7,8	16%	16	57.5	28%	OK	1
	2	5,7,8	16%	20	73.67	27%	OK	2
	3	2	16%	4	14.58	27%	OK	3
	4	6	16%	12	32.17	37%	OK	4
	A	1	–	–	27.67	–	OK	5
	B	1	–	–	32.5	–	OK	6
	C	5,6	16%	29.67	60.5	49%	OK	7
	D	5	16%	32	56.5	57%	OK	8
	E	5	16%	35.5	48.67	73%	OK	9

Notes:

1. Braced wall line 1 contains bracing segments with Methods (reading left to right): 7, 8, 2 and 2.
2. Braced wall line 2 contains bracing segments with Methods (reading left to right): 7, 5 (double sided), 8 and 5.
3. Braced wall line 3 contains a single bracing segment of Method 2.
4. Braced wall line 4 contains bracing segments with Methods (reading left to right): 6, 6 and 6.
5. Braced wall line A contains two Method 1 let-in bracing segments. Let-in bracing does not need to meet a minimum amount required, but is required at ends, or 12.5 ft from the end, and every 25 ft on center.
6. Braced wall line B contains two bracing segments, both Method 1. Braced wall line B ends at the garage.
7. Braced wall line C contains bracing segments, Method 6 and Method 5.
8. Braced wall line D contains bracing segments with Methods (reading top to bottom): 5, 5 (double sided), 5 and 5.
9. Braced wall line E contains three Method 5 bracing segments.

Example 10.5: A house in Seismic Design Category B and wind zone 105 mph. Four-foot wood structural panel bracing segments (Method 3), alternate bracing segments (R602.10.6.2), structural fiberboard bracing segments (Method 4) and gypsum bracing segments (Method 5) will be used to brace the house.

Figures 10.12a and 10.12b show the house and braced wall segments. Table 10.6 summarizes the amount of bracing required and the amount of bracing provided for each braced wall line.

FIGURE 10.12a
FIRST FLOOR PLAN WITH METHOD 3, METHOD 5 AND ALTERNATE BRACING SEGMENTS SHOWN

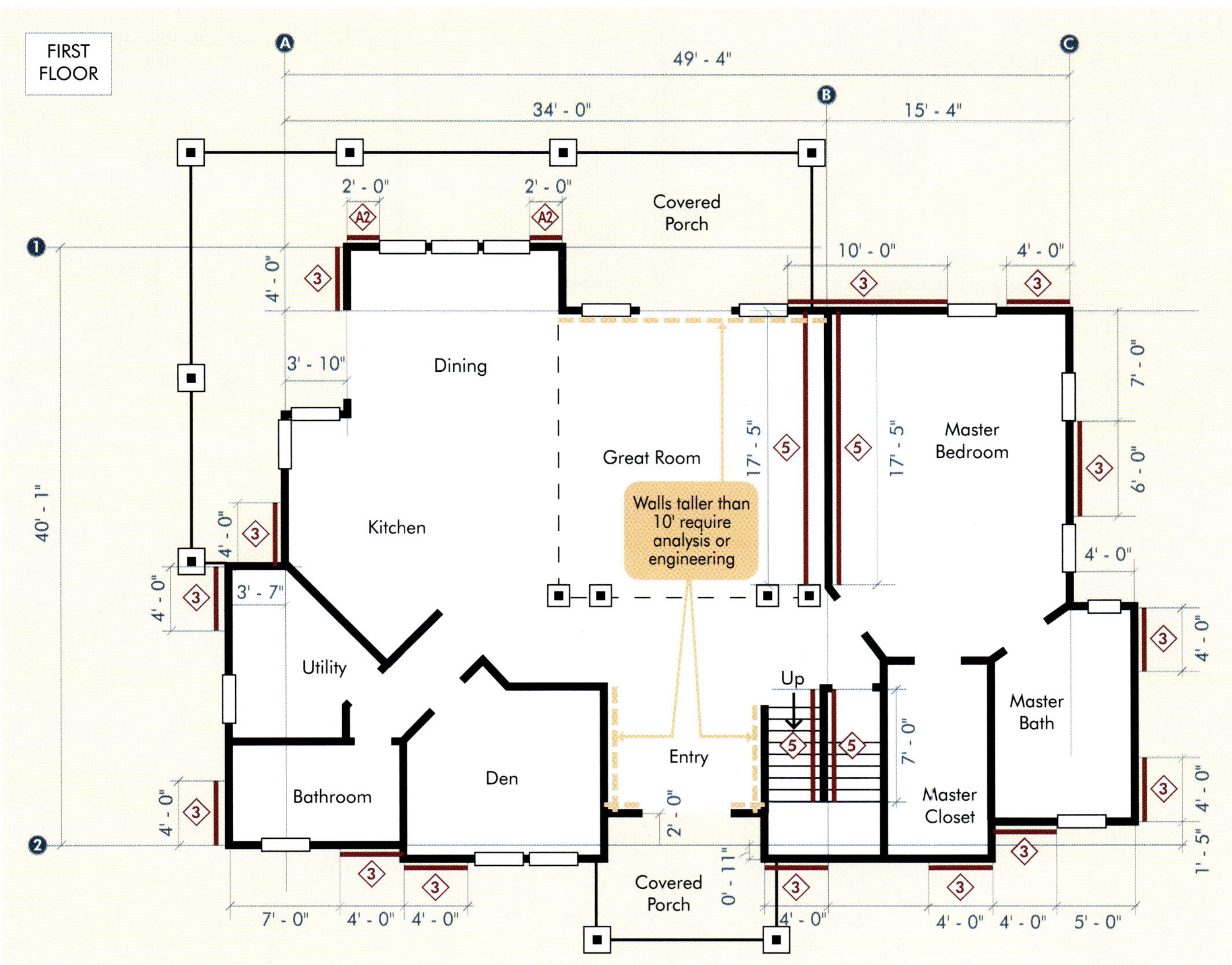

FIGURE 10.12b

SECOND FLOOR PLAN WITH METHOD 4, AND METHOD 5 BRACING SEGMENTS SHOWN

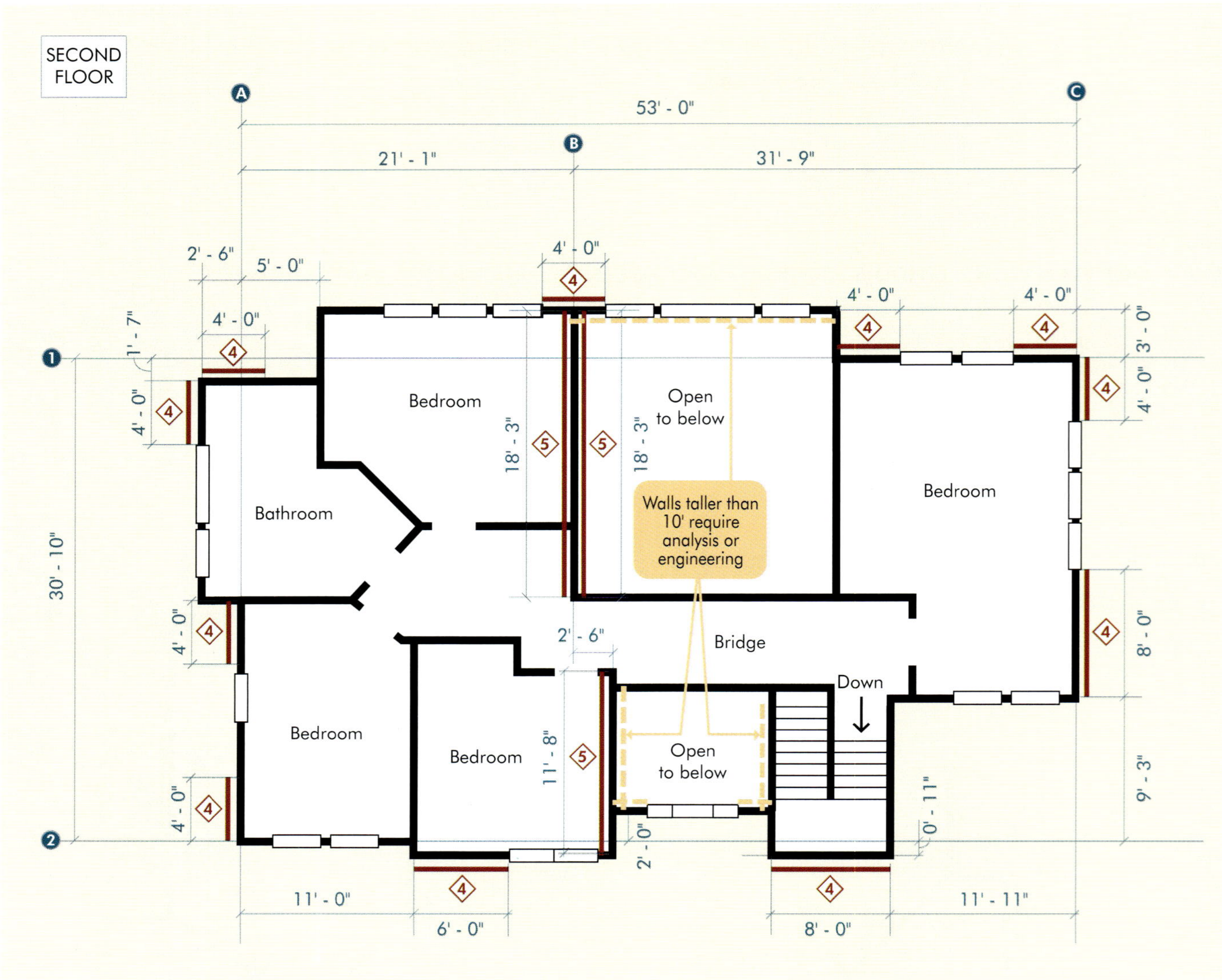

Example 10.5 highlights several important issues:

- The bracing amount required is based on 105 mph winds, which is the "greater than 100 mph but less than 110 mph" category.
- Use of the 24-inch-wide (Method A2) portal frame alternate bracing.
- Use of braced wall spacing greater than 35 feet.
- Use of walls greater than 10 feet tall.
- Use of walls having offsets that exceed 4 feet.
- Use of bracing on the second story that does not have bracing on the first story aligned below it.
- Use of various bracing methods in different wall lines and stories.
- Describes how to check the amount of bracing required to the amount of bracing present.

TABLE 10.6

BRACING REQUIREMENT FOR EXAMPLE 10.5

The amount of bracing required and the amount of bracing provided for each braced wall line on each story for Example 10.5

Story	Braced Wall Line	Bracing Method	Bracing Amount Required %	Bracing Length Provided (ft)	Braced Wall Line Length (ft)	Bracing Amount Provided %	Status	Notes
First of Two	1	A2, 3	35%	22	49.25	45%	OK	1
	2	3	35%	20	56.83	35%	OK	2
	A	3	30%	16	40.083	40%	OK	3
	B	5	45%	24.4	40.083	61%	OK	4
	C	3	30%	14	40.083	35%	OK	5
Top of Two	1	4	25%	16	55.5	29%	OK	6
	2	4	25%	14	53	26%	OK	7
	A	4	25%	12	30.83	39%	OK	8
	B	5	25%	29.9	34.75	86%	OK	9
	C	4	25%	12	30.83	39%	OK	10

Notes (designate first and second story when referencing braced wall lines per story):

1. Braced wall line 1 contains bracing segments with Methods (reading left to right): A2, A2, 3 and 3 (30% bracing required). For wall lines 1 and 2, since braced wall line spacing exceeds 35 ft, the bracing amount must be increased by 40/35 = 1.15. 1.15 x 30% = 34.5%, use 35%.
2. Braced wall line 2 contains five Method 3 bracing segments. For wall lines 1 and 2, since braced wall line spacing exceeds 35 ft, the bracing amount must be increased by 40/35 = 1.15. (See IRC Section R602.10.1.1.)
3. Braced wall line A contains four Method 3 bracing segments.
4. Braced wall line B contains two Method 5 bracing segments, both double-sided.
5. Braced wall line C contains three Method 3 bracing segments.
6. Braced wall line 1 contains four Method 4 bracing segments.
7. Braced wall line 2 contains two Method 4 bracing segments.
8. Braced wall line A contains three Method 4 bracing segments.
9. Braced wall line B contains two Method 5 bracing segments, reading top to bottom, the first double-sided and the second single-sided only.
10. Braced wall line C contains two Method 4 bracing segments.

The walls that exceed the 10-foot story height limit require engineering to show that they can resist gravity loads, combined with lateral loads, from wind pressures. Table R602.3.1 may be used to size over-height-limit studs, provided the situation meets the limitations of the table. For bracing purposes, these tall walls are not counted as contributing any bracing and are simply considered the same as an opening.

For braced wall lines 2 and C at the second story level, each braced wall line contains an offset that exceeds 4 feet. For purposes of computing the bracing present in the wall line, the offset portion is not counted as contributing any bracing. It is, however, still used to compute the total length of the braced wall line (note the length of braced wall line 2 = 53 feet, and braced wall line C = 30 feet 10 inches). Essentially, the portion of the wall that is offset greater than 4 feet is simply considered the same as an opening.

Example 10.6: A house in Seismic Design Category B and wind zone 105 mph. Continuous wood structural panel bracing segments (R602.10.5) and gypsum bracing segments (Method 5) will be used to brace the house.

Figures 10.13a and 10.13b show the house and braced wall segments. Table 10.7 summarizes the amount of bracing required and the amount of bracing provided for each braced wall line.

FIGURE 10.13a

FIRST FLOOR PLAN WITH CONTINUOUS WOOD STRUCTURAL PANEL BRACING

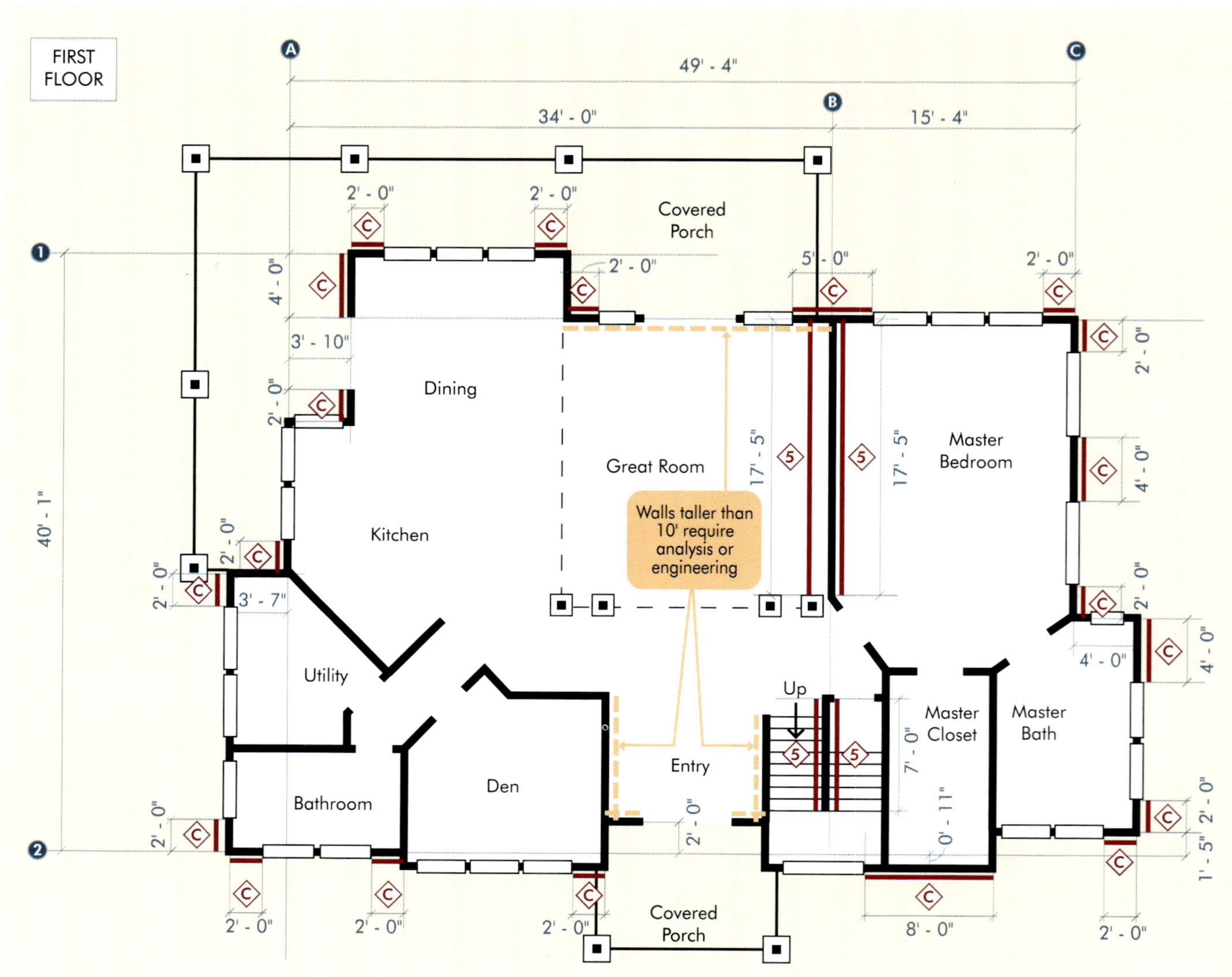

FIGURE 10.13b

SECOND FLOOR PLAN WITH CONTINUOUS WOOD STRUCTURAL PANEL BRACING

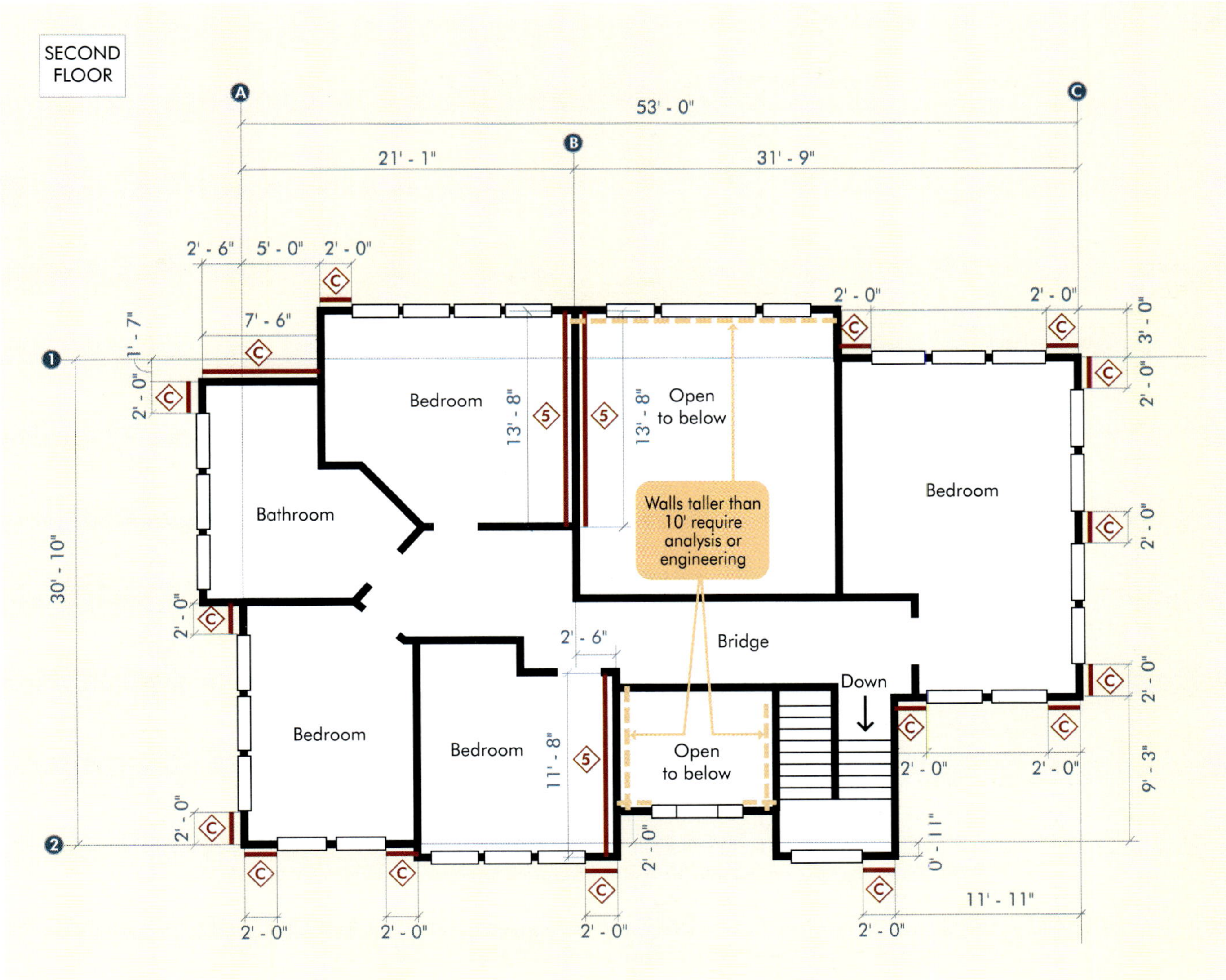

Example 10.6 highlights several important issues:

- The bracing amount required is based on 105 mph winds, which is the "greater than 100 mph but less than 110 mph" category.
- Use of continuous wood structural panel bracing.
- Use of braced wall line spacing greater than 35 feet.
- Use of walls greater than 10 feet tall. Walls other than those indicated at 10 feet are 8 feet in height.
- Bracing on the second story does not have bracing on the first story aligned below it.
- Walls having offsets that exceed 4 feet.
- Use of various bracing methods in different wall lines and stories.
- Describes how to check the amount of bracing required to the amount of bracing present.

TABLE 10.7

BRACING REQUIREMENT FOR EXAMPLE 10.6

The amount of bracing required and the amount of bracing provided for each braced wall line on each story for Example 10.6

Story	Braced Wall Line	Bracing Method	Bracing Amount Required %	Bracing Length Provided (ft)	Braced Wall Line Length (ft)	Bracing Amount Provided %	Status	Notes
First of Two	1	C	27%	13	45.497	29%	OK	1
	2	C	27%	16	56.9	28%	OK	2
	A	C	27%	12	40.083	30%	OK	3
	B	5	45%	24.4	40.083	61%	OK	4
	C	C	24%	14	40.083	35%	OK	5
Top of Two	1	C	16%	13.5	55.5	24%	OK	6
	2	C	12.8%	8	53	15%	OK	7
	A	C	12.8%	6	30.83	19%	OK	8
	B	5	25%	25.3	34.75	73%	OK	9
	C	C	12.8%	6	30.83	19%	OK	10

Notes:

1. Braced wall line 1 is fully sheathed with five full-height segments. Amount of bracing required is multiplied by a factor of 0.9 because maximum opening height is 0.85H.
2. Braced wall line 2 is fully sheathed with five full-height segments. Amount of bracing required is multiplied by a factor of 0.9 because maximum opening height is 0.85H.
3. Braced wall line A is fully sheathed with three full-height segments. Amount of bracing required is multiplied by a factor of 0.9 because maximum opening height is 0.85H.
4. Braced wall line B is sheathed with two, double-sided Method 5 segments.
5. Braced wall line C is fully sheathed with five full-height segments. Amount of bracing required is multiplied by a factor of 0.8 because maximum opening height is 0.65H.
6. Braced wall line 1 is fully sheathed with four full-height segments.
7. Braced wall line 2 is fully sheathed with four full-height segments. Amount of bracing required is multiplied by a factor of 0.8 because maximum opening height is 0.65H.
8. Braced wall line A is fully sheathed with three full-height segments. Amount of bracing required is multiplied by a factor of 0.8 because maximum opening height is 0.65H.
9. Braced wall line B is braced with two Method 5 bracing segments, from top to bottom, double-sided and single-sided.
10. Braced wall line C is fully sheathed with three full-height segments. Amount of bracing required is multiplied by a factor of 0.8 because maximum opening height is 0.65H.

The walls that exceed the 10-foot story height limit require engineering to show that they can resist gravity loads combined with lateral loads from wind pressures. Table R602.3.1 may be used to size over height limit studs provided the situation meets the limitations of the table. For bracing purposes, these tall walls are not counted as contributing any bracing and are considered the same as an opening.

For braced wall lines 2 and C at the second story level, each braced wall line contains an offset that exceeds 4 feet. For purposes of computing the bracing present in the wall line, the offset portion is not counted as contributing any bracing. The offset portion, however, is still used to compute the total length of the braced wall line (note the length of braced wall line 2 = 53 feet, and braced wall line C = 30.83 feet). Essentially, the portion of the wall that is offset greater than 4 feet is considered the same as an opening. Note also that the offset portions of the wall (not counted) have more than adequate amounts of bracing (at ends, and meet the amount required).

Can Method 5 interior braced wall lines be used with continuously sheathed Method 3 bracing?

This is another important gray area in the code. Section R602.10.5 currently requires all interior and exterior braced wall lines to be constructed of wood structural panel sheathing. The specific language "and interior braced wall lines" was added during the 2000 code cycle based on a number of floor amendments. Later analysis of the final language by users, academics and even the original proponents, led to the conclusion that the requirement for all Method 3 interior braced wall lines was not intended. The ICC Ad Hoc Committee on Wall Bracing has recommended for approval a code change proposal to eliminate this requirement from future codes. Example 10.6 represents the application of continuous wood structural panel wall sheathing with interior braced wall lines of gypsum wallboard, Method 5. In example 10.6, Method 5 bracing was used, as is normally the case for interior walls because they are already sheathed in gypsum wallboard and sufficient length is available.

Gypsum bracing

Note that a standard bracing unit for all of the panel-type bracing methods is 4 feet, except for gypsum wallboard. (See Section R602.10.4.) The strength properties for gypsum wallboard are such that it takes a 4-foot, two-sided gypsum-sheathed braced wall panel (or an 8-foot one-sided) to equal the capacity of a single one-sided, 4-foot braced wall panel of any other panel type. While not specifically addressed in the 2006 IRC, the ICC Ad Hoc Committee on Wall Bracing has proposed that, when computing percentage bracing to meet the Table R602.10.1 requirements, the percentage requirements for Method 5 wall bracing must be doubled or two-sided Method 5 walls used. (See Figure 10.14.) When such an interpretation is used, or when such language is adopted in the IRC, the examples in this chapter will have to be changed to insure that sufficient interior bracing is provided.

FIGURE 10.14

CALCULATION OF METHOD 5 PERCENT BRACING IN 2006 IRC

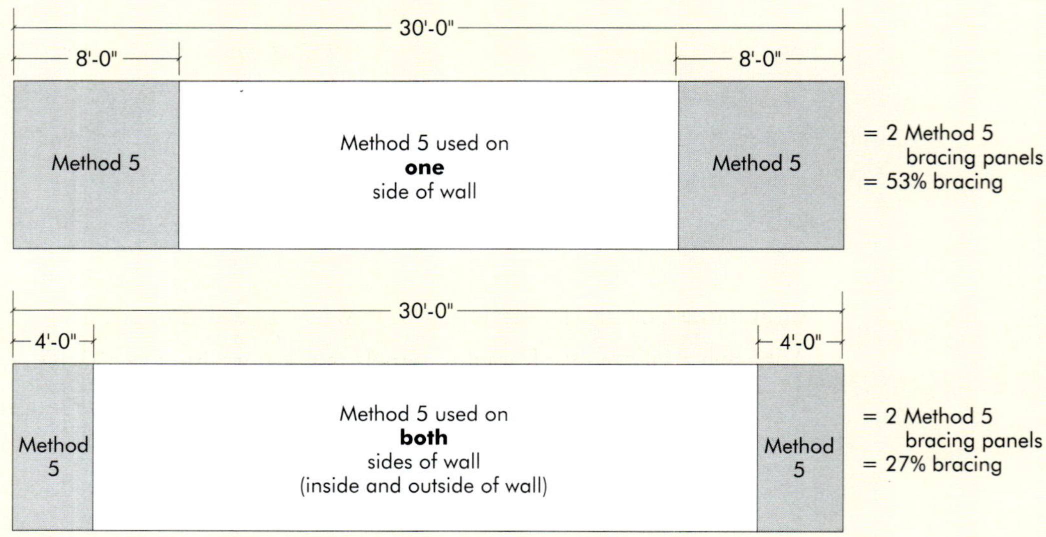

Current IRC is silent on the proper way to compute percent bracing for single- and double-sided gypsum board walls. The ICC Ad Hoc Committee on Wall Bracing has proposed changes that would make both of the above equal to 27% bracing.

What are the corner requirements for continuous wood structural panel bracing?

As previously discussed in Chapter 5, the continuous wood structural panel bracing method requires that corners be constructed in accordance with Figure R602.10.5. However, the current language in the 2006 IRC leaves room for interpretation. For example:

- No mention is made of any length requirements for the panels that are required at corners. The original testing, and intent, is to require a minimum 2-foot-long panel piece (full height of the wall) at the required corners.
- It's not clear whether the corner detail is supposed to occur at every building corner or just at the ends of continuous braced wall lines. As of this writing, the ICC Ad Hoc Committee on Wall Bracing has proposed to require the corner detail only at the ends of braced wall lines.
- Although all other bracing methods are permitted to start 12.5 feet from the end of braced wall lines, corner windows cannot currently be used with continuous wood structural panel bracing. The ICC Ad Hoc Committee on Wall Bracing has proposed for future codes the use of an 800 lb. hold down on the bracing segments at the end of the continuous braced wall line in lieu of the corner detail.

Bracing in high seismic regions

Only one example was provided for bracing in high SDCs for several reasons. Aside from the fact that significantly higher amounts of bracing are required in a braced wall line (75 percent of the braced wall line in some cases), some of the additional requirements are:

- Braced wall line spacing cannot exceed 25 feet, with one exception, as discussed in R602.10.11.1 (Chapter 4).
- Braced wall panels must occur at ends of braced wall lines with exceptions for wood structural panel bracing, as discussed in R602.10.11.2 (Chapter 7).
- Adhesive attachment of bracing panels is not permitted (R602.10.11.5).
- Minimum 3- x 3- x 0.229-inch (9 gage) steel plate washers are required on all braced wall line sill plates, as described in R602.11.1 (Chapter 8).
- Additional attachment requirements, as discussed in R602.11 (Chapter 8).
- Additional foundation requirements, as discussed in R602.10.9 and R602.2011 (Chapter 8).
- Houses cannot have irregularities, as discussed in R301.2.2.4 (Chapter 4).

These additional requirements make prescriptive bracing in high seismic regions less common than elsewhere. In many high seismic regions, the lateral-force-resisting system is engineered in most houses, primarily because modern home designs have features or irregularities which exceed the prescriptive IRC limits.

A very good resource for wall bracing in high seismic regions is the *Homebuilders Guide to Earthquake Resistant Design and Construction* (FEMA 232).

Regional differences

How IRC bracing is viewed differs substantially by region. In many high seismic regions, the bracing requirements are perceived as being more liberal because engineers and builders are accustomed to more robust lateral force resisting systems with detailed load paths, which include the use of hold-down devices. In low seismic regions, where houses are typically not engineered, the IRC wall bracing requirements are sometimes viewed as too conservative (too stringent). These opposing views are one of the reasons why wall bracing provisions seem to be constantly changing. They are continually being modified by both ends of the spectrum.

BIBLIOGRAPHY

AF&PA. 2005. National Design Specification for Wood Construction. American Forest & Paper Association American Wood Council, Washington, DC.

AF&PA. 2001. Wood Frame Construction Manual. American Forest & Paper Association American Wood Council, Washington, DC.

AISI. 2001. Standard for Cold-Formed Steel Framing – Prescriptive Method for One-and Two-Family Dwellings (COFS/PM) with Supplement to Standard for Cold-Formed Steel Framing - Prescriptive Method for One-and Two-Family Dwellings. American Iron and Steel Institute (AISI), Washington, DC.

APA. 2003. Narrow Garage Wall Bracing for One- and Two-Story Homes. Form TT-077. APA – The Engineered Wood Association, Tacoma, WA.

APA. 2003. IRC Continuous Structural Panel Sheathing Option to Meet Wall Bracing Requirements. Form TT-079. APA – The Engineered Wood Association, Tacoma, WA.

APA. 2005. Collector Design for Bracing in Conventional Construction. Form TT-102. APA – The Engineered Wood Association, Tacoma, WA.

APA. 2005. Introduction to Wall Bracing. Form F430. APA – The Engineered Wood Association, Tacoma, WA.

APA. 2006. Hurricane Katrina – Structural Performance of Wood Frame Buildings in the Aftermath. Form SPE-1125. APA – The Engineered Wood Association, Tacoma, WA.

ASCE. 2005. Minimum Design Loads for Buildings and Other Structures. ASCE-7-05. American Society of Civil Engineers, Reston, VA.

FEMA. 2006. Homebuilders Guide to Earthquake Resistant Design and Construction. FEMA 232. Federal Emergency Management Administration, Jessup, MD.

IBC. 2006. International Building Code. International Code Council, Falls Church, VA.

IO&TFDC. 1998. International One- and Two-Family Dwelling Code. International Code Council, Falls Church, VA.

IRC. 2006. International Residential Building Code. International Code Council, Falls Church, VA.

IRC. 2004. Supplement to the 2003 International Residential Code. International Code Council, Falls Church, VA.

Keith, E., Z. Martin, T. Skaggs, B. Yeh. 2006. The Evolution of Lateral Load Design in Residential Construction. Proceedings, 2006 Timber Conference, Portland, OR.

NIST Technical Note 1476. Performance of Physical Structures in Hurricane Katrina and Hurricane Rita: A Reconnaissance Report. National Institute of Standards and Technology, Gaithersburg, MD.

O&TFDC. 1971. CABO One- and Two-Family Dwelling Code. Council of American Building Officials, Falls Church, VA.

O&TFDC. 1986. CABO One- and Two-Family Dwelling Code. Council of American Building Officials, Falls Church, VA.

O&TFDC. 1995. CABO One- and Two-Family Dwelling Code. Council of American Building Officials, Falls Church, VA.

SBCCI. 1999. Standard for Hurricane Resistant Residential Construction. SSTD-10. Southern Building Code Congress International, Birmingham, AL.

van de Lindt, J.W., A. Graettinger, R. Gupta, S. Prior, T. Skaggs, K. Fridley, 2006. Damage Assessment of Woodframe Residential Structures in the Wake of Hurricane Katrina. Report to National Science Foundation.

van de Lindt, J.W., A. Graettinger, R. Gupta, S. Prior, T. Skaggs, K. Fridley. 2007. Performance of Woodframe Structures During Hurricane Katrina. ASCE Journal of Performance of Constructed Facilities. To appear 21(2).

Residential Wall Bracing Principles and Design Options

by Jay Crandell, P.E.; and Susan Herrenbruck

The *International Residential Code* (IRC) requires conventional wood-frame dwellings to be braced against lateral (racking) forces associated with high winds and earthquakes. To achieve this critical structural safety objective, several wall bracing options are provided in IRC Section R602.10. By promoting a sound understanding and application of the IRC bracing provisions, designers, builders and code officials can work together toward the construction of code-compliant dwellings for a diverse housing market.

The proper design of code-compliant wall bracing includes several steps and may allow for multiple solutions—some of which may not work for all walls on a given building. This article addresses five principles in order to help eliminate possible confusion and provide direction on important design considerations and options. Additional information on wall bracing and a more thorough explanation of these concepts and design principles, as well as supplemental design data, can be found in the *Residential Structural Design Guide*, which is available free of charge from the U.S. Department of Housing and Urban Development at *www.huduser.org*.

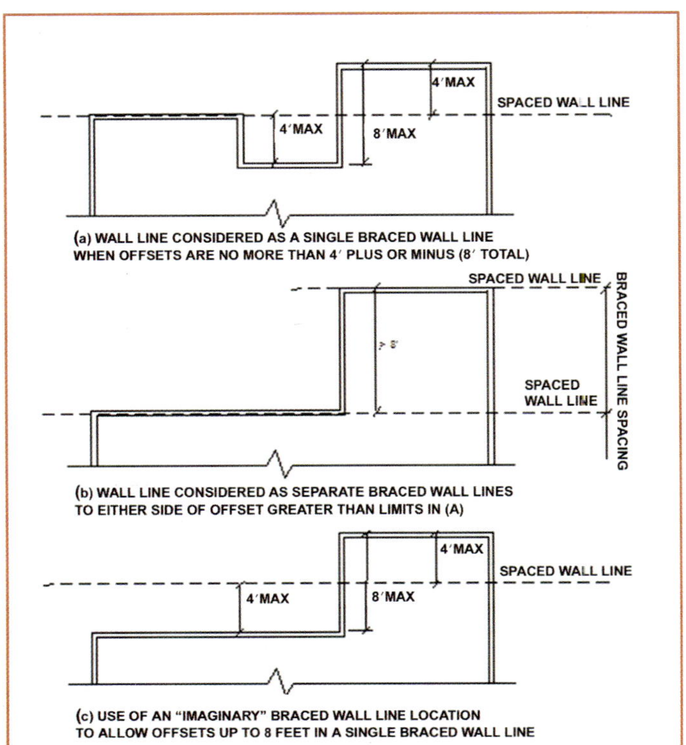

Braced wall line offsets.

Wall Bracing Concepts and Requirements

The following concepts and requirements are fundamental to the wall bracing provisions of IRC Section R602.10. They can also potentially impact the architectural design of individual building walls as well as the layout plan of all exterior walls.

Braced Wall Line (Section R602.10.1)

Walls that are braced to resist racking are known as braced wall lines. Essentially, all exterior walls are considered to be braced wall lines and are required to be properly braced with braced wall panels. Although not always required, interior walls may also be used as braced wall lines.

A braced wall line can have limited offsets and still be considered as a single braced wall line. When an offset does not comply with those limits, the wall lines to either side of the offset are required to be considered as separate braced wall lines. Such offsets affect the amount and location of braced wall panels and, thus, the layout of openings for windows and doors in a series of braced wall lines on a given building side (elevation).

Braced Wall Line Spacing (Section R602.10.1.1)

Braced wall line spacing determines the amount of racking load that is tributary to the two or more parallel braced wall lines in each plan direction. The racking load must be resisted by use of an adequate amount of braced wall panels on each braced wall line. Therefore, bracing amounts are dependent on the spacing of braced wall lines.

This consideration influences the space that is available for wall openings on exterior walls, which may necessitate

needed. Rather, as we have known all along, survivable space must be provided between structures in the interface and highly flammable vegetation and the structures must have an appropriate level of ignition resistance. This does not mean that homes must be surrounded by concrete or some other noncombustible material, only that landscaping should employ plants with a high moisture content and that any other vegetation which may place the building at risk should be trimmed or replaced with high moisture content plants.

Given the consensus on how to address the problem, why are we still losing thousands of structures every year and risking thousands of lives? Consider for a moment another low-cost, highly effective, common-sense safety measure: the use of seatbelts in automobiles. In the 1960s, the federal government mandated that seatbelts be installed by automobile manufacturers and launched a robust public education campaign. More people began using seatbelts, but not nearly enough to achieve the projected national safety benefits. It was only after the federal government tied state highway funding to mandatory seatbelt laws that their use became the rule rather than the exception, resulting in a substantial reduction of highway injuries and deaths.

The current situation with the wildland-urban interface is similar. We've seen time and again how at-risk neighborhoods and even individual owners that adopt the recommended measures sustain little or no damage from wildfires while adjacent properties are utterly destroyed. So although we should certainly continue our education efforts, it is time that we recognize that they will only be minimally effective in the absence of appropriate regulations.

The *International Wildland Urban Interface Code* (IWUIC) contains all of the necessary regulatory language as well as a number of instructional appendices to assist in the application of its provisions. Of course, it is also fully compatible with the *International Building Code, International Residential Code* and *International Fire Code*: the most widely used model codes of their type in the U.S. There can be no doubt that the adoption and enforcement of the IWUIC in interface areas will create a level of wildfire safety never before realized in this country—and with development in these areas continuing at an unprecedented pace, the time to press for such action is now. ◆

the use of interior braced wall lines to ...lp share the bracing load. It also ...s the amount of bracing required ... braced wall line spacing ... ...mum limits prescribed in ...

...all Panel (...R602.10.1)

...nel is a section of a braced ...s braced via a code-compliant ...raced wall panels must meet ...n width requirements to count toward ...ed bracing amounts. The minimum widths required for braced wall panels using the various allowed bracing methods constrain the layout and spacing of wall openings in a code-compliant braced wall line. While braced wall panels that are narrower than allowed contribute to bracing, they currently are not taken into consideration.

Braced Wall Panel Location (Section R602.10.1)

In addition to being used to meet minimum bracing amounts, the location of braced wall panels on braced wall lines must meet two additional requirements:

- they must be spaced no greater than 25 feet on center along a braced wall line; and

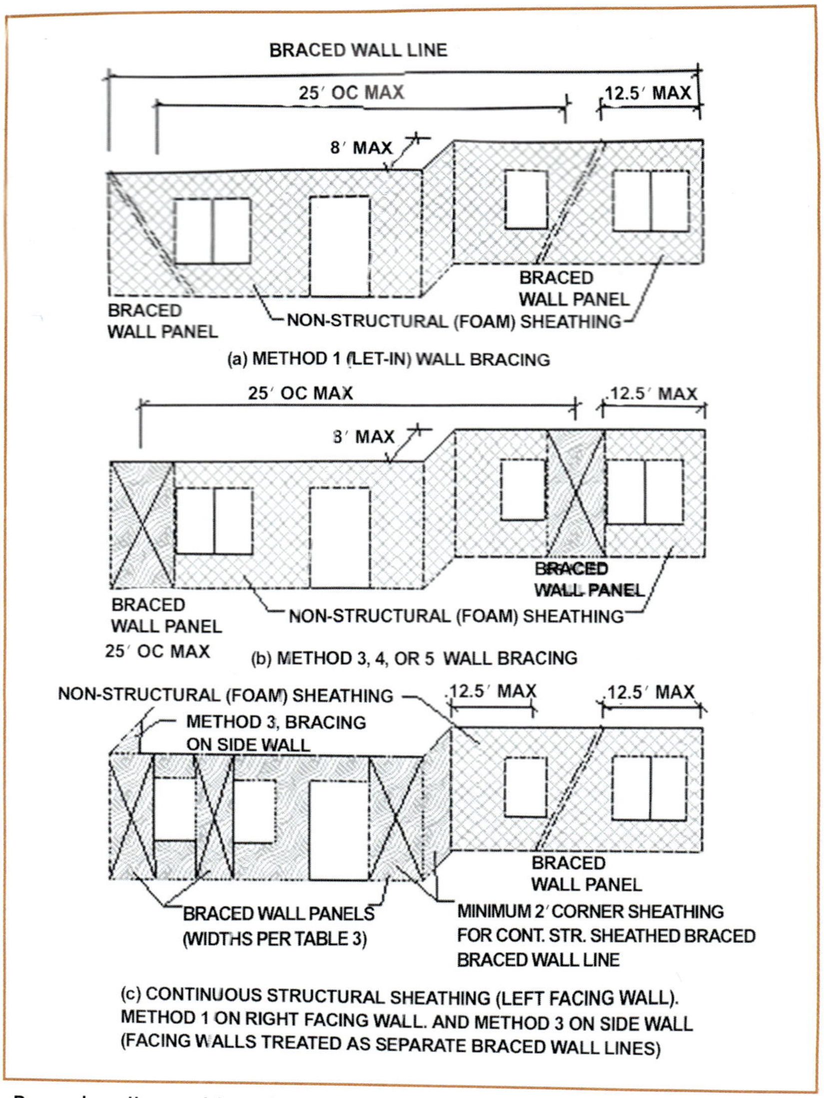

Braced wall panel location.

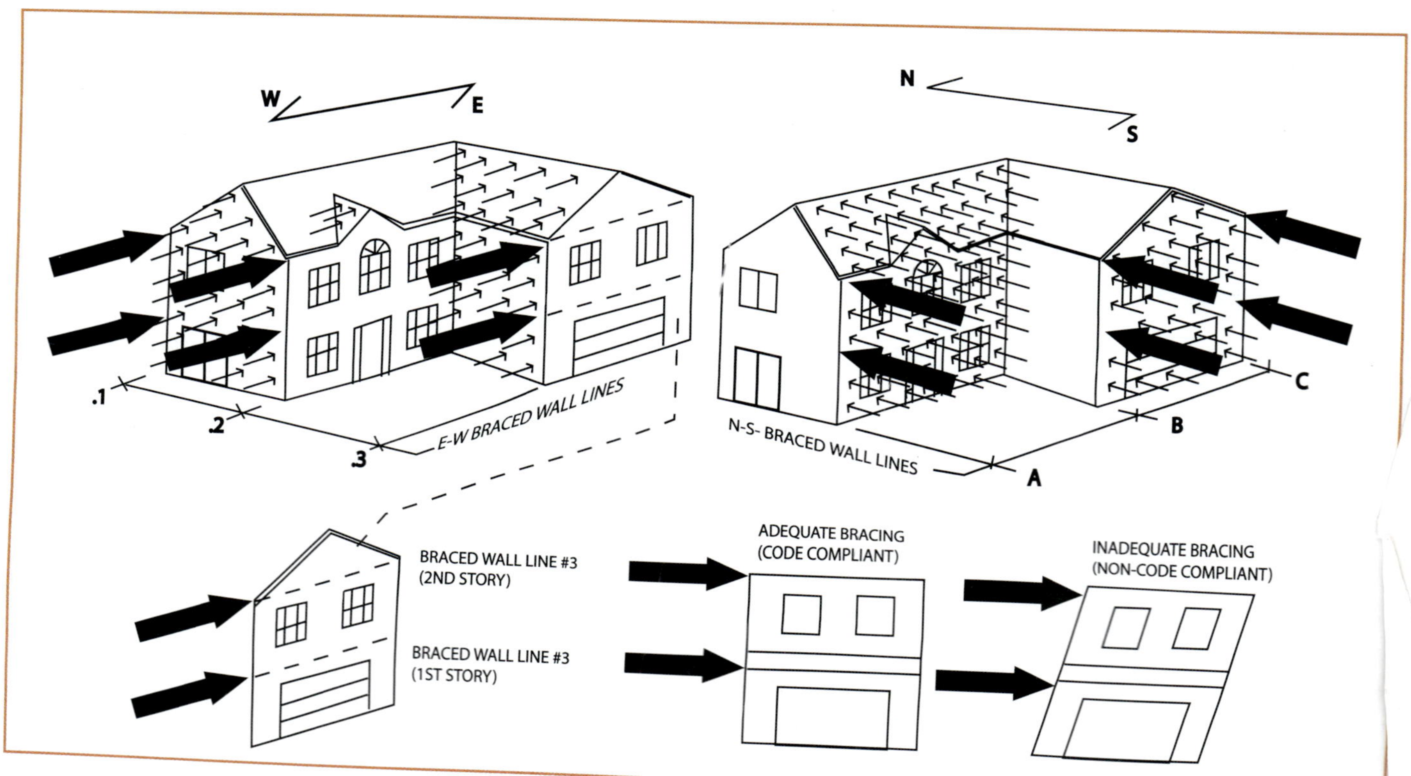

Braced wall wind loads.

- in most (low seismic risk) areas of the country, they must begin no more than 12.5 feet from the end of a braced wall line—usually defined as an inside or outside building corner.

Use of Multiple Bracing Methods

The compatibility of the various allowed bracing methods is not directly addressed in the IRC. However, the code recognizes specific instances in which one brace type can be substituted for another (see, for example, Section R602.10.6). Because each code-compliant bracing method is intended to provide equivalent minimum performance, they may be used interchangeably on different braced wall lines of the same building or within the same braced wall line provided that each bracing method complies with its specific requirements. This concept is important because it allows bracing methods to be selected and used on the basis of conditions that may be unique to each braced [illegible] thereby allowing the overall brac[illegible] be optimized for cost and possible [illegible] for nonstructural considerations such as en[illegible] building.

Table 1. Functions, code requirements and typical materials for foam-sheathed wall systems.

FUNCTION	WALL COMPONENT	2003 IRC REFERENCE	TYPICAL MATERIALS
support of gravity loads	wall frame (studs, plates, headers, etc.)	R602.1 through R602.7	2 x 6 studs 2 x or engineered headers
racking resistance (bracing)	various materials and methods	R602.10	1 x 4 let-in braces or equivalent metal braces, wood structural panels, fiberboard, gypsum board, etc.
weather resistance	weather-resistant barrier	R703.2	approved water-resistant foam sheathing
	cladding	R703	vinyl siding, wood siding, cement composite siding, etc.
cladding support	siding backer board	R314.2.5	minimum ½-inch-thick foam sheathing (siding attached to studs per manufacturer requirements and IRC Table R703.4)
energy efficiency	cavity insulation	R314, R316, Chapter 11	fiberglass batts
	exterior insulation		½-inch to 1-inch foam sheathing
moisture vapor resistence	vapor retarder (as required)	R318	Kraft paper, polyethylene, etc.

Wall Bracing Design Approach

The following steps outline a routine approach for proper[illegible] applying the IRC's wall bracing provisions.

- **STEP 1.** Identify braced wall lines on a preliminary plan (e.g., check offsets and label separate braced wall lines on each side of the building).
- **STEP 2.** For each plan direction (North–South and East–West) and story level, determine the spacing between braced wall lines.
- **STEP 3.** Select a bracing method for each braced wall line and determine the required bracing amounts for each braced wall line.
- **STEP 4**. Determine the bracing amount provided by code-compliant braced wall panels on each braced wall line and verify that the minimum required bracing amount from Step 3 is met or exceeded.
- **STEP 5.** Verify braced wall panel spacing limit (e.g., maximum 25 feet on center) and corner distance (e.g., braced wall panel starts no more than 12.5 feet from the end of a braced wall line) are met for each braced wall line. If the bracing requirements are not met in Steps 4 or 5, adjustments to the bracing technique or wall configuration will be necessary to achieve compliance.

Optimizing Wall Design

A single bracing method may not solve all of the desired wall functionality in every building design. As illustrated in Table 1, only one of the six functions of a wall system is related to bracing. However, bracing materials can affect other important wall functions. For example, Section R602.10.5 of the 2006 IRC requires that an entire building be continuously sheathed with wood structural panels, even though in typical cases only braced wall lines with narrower than 48-inch-long braced wall panels necessitate the use of this method. As a consequence, 2 x 6 framing may be required in order to meet energy code requirements for cavity insulation.

One alternative, foam sheathing, serves multiple functions including insulating for energy code compliance while simultaneously creating a weather-resistant barrier and is sometimes used as a backer board for exterior finishes. With proper consideration of bracing, foam-sheathed wall systems may allow the use of standard 2 x 4 construction while still